ACHIEVE LASTING PROCESS IMPROVEMENT

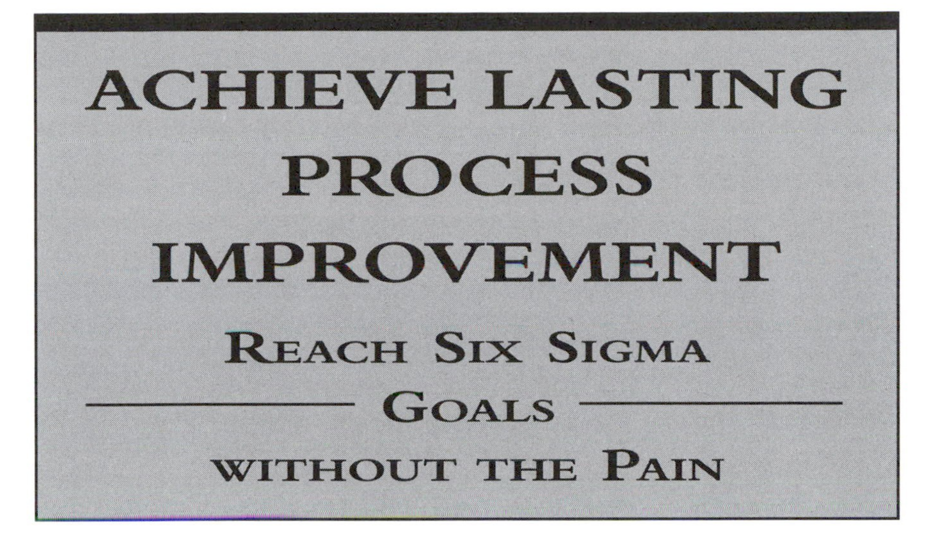

Bennet P. Lientz
Anderson Graduate
School of Management
University of California,
Los Angeles

Kathryn P. Rea
The Consulting Edge, Inc.
Los Angeles, California

Amsterdam Boston London New York Oxford Paris
San Diego San Francisco Singapore Sydney Tokyo

This book is printed on acid-free paper. ∞

Academic Press
An imprint of Elsevier Science.
525 B Street, Suite 1900, San Diego, California 92101-4495, USA
http://www.academicpress.com

Academic Press
32 Jamestown Road, London NW1 7BY, UK
http://www.academicpress.com

Library of Congress Catalog Card Number: 2001099786

International Standard Book Number: 0-12-449984-8

PRINTED IN THE UNITED STATES OF AMERICA
02 03 04 05 06 07 MM 9 8 7 6 5 4 3 2 1

CONTENTS

PART I
PREPARE FOR PROCESS IMPROVEMENT

CHAPTER 1
Introduction

CHAPTER 2
Understand Your Business

CHAPTER 3
Assess Technology and Industry Factors

PART II
DEFINE THE NEW BUSINESS PROCESSES

CHAPTER 4
Select the Right Processes

CHAPTER 7
Define Your New Business Processes and Quick Wins

PART III
PLAN FOR PROCESS IMPLEMENTATION

CHAPTER 8
Develop Your Improvement Implementation Strategy

CHAPTER 9
Develop Your Improvement Implementation Plan

PART IV
IMPLEMENT AND SUSTAIN PROCESS IMPROVEMENT

CHAPTER 10
Accomplish Quick Wins

CHAPTER 11
Implement Improved Processes

CHAPTER 12
Measure and Maintain Improvement Momentum

PART V
DEAL WITH SPECIFIC ISSUES

CHAPTER 13
Management Issues

CHAPTER 14
Business and Organizational Issues

CHAPTER 15
Work Issues

CHAPTER 16
Political and Cultural Issues

PREFACE

METHODS FOR PROCESS IMPROVEMENT AND WHY THEY HAVE FAILED

A business process is a combination of procedures, systems, policies, and organizational elements that performs a core function of a business. For thousands of years, organizations have depended on their business processes for survival and growth. Business processes are responsible for all profits and losses. Obviously, people have been trying for centuries to get the processes right. Records of this effort go back to ancient Roman, Chinese, and Egyptian times. We are still at it today.

Over the years, there have been many attempts to find methods to improve business processes. These methods have sometimes been focused on how the work is done, how the work is managed, how the work is automated, and on who performs the work. Examples have included the following:

- *Industrial engineering,* which attempted to implement small improvements to processes and measure processes. The contribution of industrial engineering has been to demonstrate the need for measurement. In many cases, the people went back to the old ways after the engineers had left.
- *Total quality management* (TQM), which aims at the widespread involvement of employees and managers to improve the quality of output from processes. TQM gave attention to the need to look at quality factors in a process and to reinforce measurement.
- *Reengineering,* which aimed at major structural change and the upheaval of business processes. The contribution of reengineering was to show that major change was feasible and desirable over incremental change.
- *Continuous process improvement,* which aimed at sustained improvements through incremental efforts. The contribution of this method was to put the spotlight on the need to implement longer lasting change.
- *Automation,* which aims to automate the business process through client-server, data warehouse, Enterprise Resource Planning (ERP), intranets/Internet-based systems.
- *Six Sigma,* which expands on TQM and focuses on the customer, measurement, proactive management, and a drive for perfection.
- *E-business,* which is a natural extension of internal automation to customers, employees, and suppliers.

BUSINESS PROCESS OBJECTIVES

All of these techniques have had instances of success, which led to their popularity. After gaining some popularity, a method often falls out of favor and popularity. Some methods have failed often. Surveys indicate a failure rate of over 50% for many of these methods. Why does this happen? Why do methods such as these fail? To answer these questions, you must first turn to the objectives related to improving processes. Here are some objectives:

- A process must be *flexible* to accommodate changing conditions.
- A process must be *efficient* so that work can be done in a cost-effective manner.
- A process must be *scalable* so that it can handle the peaks and troughs of work without falling apart.
- A process must be *effective* so that it can handle any type of work without creating exceptions, workarounds, and new systems that arise to address individual situations.
- A process must be *situation dependent* so that it can be performed in a variety of countries, cultures, political and legal structures, companies, and so on. This objective, of course, links to flexibility.
- A process must be *measurable* so that people can determine the status of a business process.

OBJECTIVES OF PROCESS IMPROVEMENT METHODS

Let us now turn to the objectives of any method or technique for process improvement:

- A method needs to be able to be applied by *organizations of any size or type.*
- A method must be able to be employed in just about *any culture.*
- A method must be *self-sustaining* so that process improvement will keep going on after the managers at the top have departed.
- A method must be *flexible* to take advantage of new technology and automation.
- A successful method must be in the best *self-interest* of people at all levels of an organization.
- A successful method must have *broad support and understanding* throughout an organization.
- A method must be *multidimensional* to include measurement, change, management, technology, methods and procedures, and policies.

- A method must be *affordable* to implement in a reasonable amount of time.
- The method must be sensitive to and take advantage of the *political realities* of the organization.

WHY METHODS FAIL

In the light of the many failures cited in the literature, it is tempting to grasp at the concept that we just have to keep searching for a silver bullet method. Ah, if we can just come up with another buzzword, the right jargon, and some examples of success, we can make it all right. This is both naive and dangerous. Process improvement methods have been known to bankrupt and drive organizations to the brink of extinction. If some jargon method appears, remember what Bette Davis said in a movie, "Fasten your seat belt, it's going to be a bumpy ride." Yet we have ignored the successful lessons of history. Take, for example, the Roman construction of roads, buildings, aqueducts, and bridges. Romans constructed and maintained these structures successfully for almost 1,000 years. How did they do it? How did they achieve such successful methods without jargon and computers? The answer was a combination of common sense and jargon-free techniques that were refined and improved over the centuries. There are many other examples—government structure in China, societal structure in Egypt, and so on.

Now go back to the list that was presented earlier. How do these methods stack up in light of the stated objectives. Unfortunately, not very well. Here are some reasons for failure.

- *Jargon dependence.* The method was constructed around buzzwords. Such a method has widespread appeal to consultants and academics who can leverage off the concept to make a lot of money. Some companies survive by moving from one hot method to the next—just like a surfer who catches successive waves. Jargon methods are doomed to fail for many reasons. First, they require constant reinforcement since they are often not intuitive. Second, they tend to be expensive and rely on outsiders.
- *Overdependence on support of top management.* Upper management support is obviously critical for the success of any method. But detailed involvement? Give us a break! Top managers have more to do than to focus much of their time on some method. They typically give it initial attention and then move on to other work. Because the method depended so much on management involvement, it starts to shrivel and decline.
- *Overreliance on outside help.* While it is probably necessary to have some outside involvement when methods are first initiated, a company just

cannot endure continued meddling by outsiders. Morale sinks. Management authority is questioned. Employees begin to ask, "Who is in charge? The managers or the consultants?"

- *One dimensional methods.* Many of the techniques mentioned have are unidimensional. They focus on just one aspect such as measurement, the customer, or automation.
- *Lack of attention to getting near-term results.* If you just focus on the longer term, you could be dead in the short term. You must get rapid results to maintain and build momentum.
- *Lack of attention to the maintenance and evolution of the process after the improvement has been implemented.* Almost all of the methods fail to address process maintenance. It is like there are missing chapters. There is the assumption that if you "fix it" it will stay fixed. Not likely. In many cases, a process will revert back to its earlier state. After all, the old process is what people have been doing for many years. They have only been doing the new process for months.

Yet, for all of the negatives of the methods, they all have tried to seek good results. This book is about how to achieve the same goals as Six Sigma and other methods, but to do so by using more realistic and direct methods. The methods in this book have been applied to more than 75 organizations on five continents. At the heart of the techniques is common sense. These are proven methods derived from techniques that worked more than 2,000 years ago. They have been enhanced and updated by the knowledge and technology acquired since then. But they are still common sense.

THE APPROACH IN THIS BOOK

Here are some critical success factors for the techniques in this book.

- Appeal to the self-interest of people at all levels of the organization.
- Grassroots involvement and support.
- Support but not intensive involvement by management.
- Use of systems and technology in a appropriate and suitable way.
- Self-sustaining measurement.
- Discouragement of exceptions and workarounds; structure for shadow systems. A shadow system is a system that a department creates to meet its individual needs. Shadow systems will not go away. Instead, they must be made to fit within the context of the business process.

What is the method called? How about "common sense"? What are the key points of the method? Here are some:

- An overall process plan for a business process
- A focus on a group of related business processes rather than a single process
- A drive to achieve near-term, tangible results
- The initial support of top management, followed by selective targeted involvement
- Widespread involvement at the lower levels of the organizations
- Measurement of the business processes from the start
- An understanding of external industry, societal, and technological factors that can impact the processes
- A modern project and program management method that involves a focus on risk management, collaboration, and lessons learned
- Gathering and using lessons learned to help sustain improvements in processes
- Maintenance and enhancement of business processes

OBJECTIVES OF THE BOOK

The book presents very specific objectives for you as you learn about process improvement. Organizations that have applied this methods have met the objectives listed:

- You can employ the techniques right away.
- You do not have to spend money to use these techniques.
- You can do a lot of the work by yourselves without major help.
- You can achieve results in the short term.
- Employees will become more motivated to achieve change.
- You can measure your work as you go.

ORGANIZATION OF THE BOOK

Most chapters are organized around the following sections:

- *Introduction.* This is the background and road map.
- *Objectives.* What should you be able to achieve with the techniques presented in this this chapter?
- *Methods.* This is the detailed approach for the topics covered.
- *Examples.* Three real-world examples are covered throughout the book.
- *Lessons learned.* These are detailed guidelines on how to employ the methods effectively and efficiently.
- *Problems you might encounter.* Some of the common problems that you might face in doing the work are discussed.

- *What to do next.* These are specific actions you can take now.

In addition there is a section at the end of the book that addresses specific management, work, business, and cultural/political factors that you are likely to encounter. To help you with these issues, the following questions are answered:

- How does the issue arise? How do you detect it?
- What is the impact of the issue?
- How can you prevent it?
- How can you resolve it if it appears?

PART I

Prepare for Process Improvement

CHAPTER 1

Introduction

WHAT ARE BUSINESS PROCESSES?

A business process is a set of related activities that produces specific end products. One example is a payroll process that produces payments to employees and reports. Another might be research and development that yields improvements over current products or services. The list is endless. Processes are all around you. You exist and participate in various processes at home and work. Processes are encompassing. A business activity or process includes the following:

- Business procedures for standard work or transactions.
- Exception transactions. These are infrequent transactions.
- Workarounds to handle specific problems in processes or where a process has broken down.
- Automation provided by the information technology (IT) department or outsourcing firms.
- Automation developed internally within departments for their own needs. You have seen many of these. Many departments are highly dependent on them for survival. In fact, they are so important in process improvement, they will be called "shadow systems."
- Policies. Policies are rules that govern the business processes and procedures. Policies are very important in this book because sometimes the shortest path to improvement is to modify or amend a specific process.

- Employees, consultants, and contractors. Most processes are either manual or partially automated. Even those that are highly automated require employees.
- Managers and supervisors. Obviously, work requires supervision and management.
- Facilities and other resources. Work does not exist in a vacuum. It must be performed in one or more locations.
- Politics and political factors. Most processes involve human beings and organizations that are by nature political. We cannot ignore this reality.

Here is a lesson learned. When you are considering improvement, don't get carried away by any one of these. Unfortunately, you have to consider them all.

Processes can be either informal or formal, or a combination of both. You deal with groups of processes all of the time. Processes very seldom exist in isolation. They are interdependent and interrelated. The same people or systems may perform different transactions from several processes. That is why you should consider groups of processes, not a process in isolation, when you do improvement. Otherwise, what you gain from one process may be taken away by inefficiencies introduced in surrounding processes.

WHAT ARE CHARACTERISTICS OF PROCESSES?

Here are some basic characteristics about business processes along with some observations.

• A company or agency survives and is successful because of its business processes. You can have a wonderful organization and superb systems, but if you don't get the processes right, you are screwed. Many dot-com failures attest to this. Yet it is absolutely amazing to us that organizations pay so little attention to them and take them for granted.

• A process deteriorates over time, if not maintained. This sounds strange at first. But a process is like a building or automobile. You have to give it attention and maintain and enhance it. What can happen to a process over time? New situations arise. These don't fit the standard procedures. So employees create exceptions. Over time, the number of exceptions, workarounds, and shadow systems may outnumber the standard transactions. People may have been trained at one time in process and system, but new employees enter the scene from time to time. Often, they are not trained. Instead, they are asked if they have done similar work someplace else. If they have, they may receive little training. They may just be thrown into the process. They may then apply what they did in their last job. In one bank we are familiar with, a service center hired three shift managers for the 24-hour operation from different banks. Each manager applied what he or she had used from a previous job! The methods were counterproductive and inconsistent. Chaos reigned for some time.

• You can fix a process, but unless the employees at the bottom (not the top) are committed to the procedures, policies, and so on, there is a major danger and risk that the process will revert back to what was there before. Automation can help here by providing more formal methods, but there can still be a backward drift.

• Processes are interdependent, as was stated earlier. So when you change something in one place, impacts may surface elsewhere.

• Processes are political—you just cannot avoid it. No matter how much you want to employ rigorous statistics, mathematics, and IT, processes involve people. A number of employees derive their power and satisfaction from their knowledge and influence in a process. When you try to change a process, you affect the power structure in departments, which is very threatening. It can create many enemies, even among employees who outwardly support change. Politics will be at the forefront of what we explore in this book.

• The people who matter in the processes are not the ones at the top, but at the bottom. It is the person on the assembly line, the teller in the bank, the customer service representative. If you use a method that does not address their fears and aspirations, your efforts will likely be doomed to failure.

• People who perform the work develop shortcuts and guidelines to help them in their work, make it more interesting, and be more productive. These will be called lessons learned. A major focus of the approach in this book is to generate lessons learned and then to use these to improve the work.

WHAT IS PROCESS IMPROVEMENT?

By definition, when you improve a business process, you make it better. But what is better? Better for whom? You want to improve business processes for everyone—employees, customers, suppliers, and managers—not just management. Otherwise, you are not likely to get lasting improvement since you ignored a critical audience and their self-interests.

Process improvement can consist of small steps at improvement or major change. It can involve some or all of the elements listed earlier as parts of a process. The approach here is to generate both short- and longer-term improvement.

PROCESS IMPROVEMENT HAS A LONG HISTORY

Think back to the history classes that you've taken. You learned about the massive construction projects of the Egyptians, Romans, Chinese, Incas, and others. Traditional teaching focused on the organization and technical methods for building pyramids, roads, aqueducts, temples, and other structures. But they left out probably the most important part. All of these civilizations had processes in place that in many cases lasted hundreds of years.

Let's consider Roman aqueducts. Once a town was sited, a Roman team of engineers showed up and planned the aqueduct. In the Roman Empire there were literally hundreds of these that had to be built and maintained. Often, they were many miles in length. The construction was very complex. How did they do it? Existing records are not very good here. The ancient Romans, like us, took the work for granted. There are some records to indicate that formal training and procedures were employed. Team members were rotated among different sites to share knowledge and experience. The success of this effort is still evident. Some of the aqueducts are still in use today—2,000 years later!

What can you learn from the successes of the past?

- Process improvement and attention to processes were an integral part in the success and sustenance of major civilizations.
- Jargon and buzzwords were not used. Instead, the actions taken were based on common sense.

That is what this book is about—common sense.

WHAT ARE THE GOALS OF PROCESS IMPROVEMENT?

It is useful to begin with general objectives and then move on to more specific goals. First, here are some general objectives:

- Processes are measured on a sustained basis. Measurement of processes makes people aware of where improvements are needed. However, the political by-product is that management and employees become much more aware of processes and business activities.
- Processes are improved and maintained by people when it is in their self-interests to do so. Mandating and supporting improvement by management is important, but this can only go so far. People have to want to do it themselves. After all, management cannot watch every process every day.
- Processes must yield benefits to all of the stakeholders and people involved or supporting the processes.

Now let's consider specific goals:

- Culture improvement. The climate in an organization must give importance and sustain processes.

- Customer satisfaction and service. Processes must provide exemplary service to customers.
- Market growth.
- Facilitate lessons learned and experience for cumulative improvement.
- Flexibility in techniques so that the number and type of exceptions is minimized.
- Quality improvement.
- New products and services. Processes must not only develop these, but support them.
- Reduced time and greater efficiency. Processes must handle the workload in a timely manner.
- Manageability and measurement. Processes must be able to be managed and measured with a reasonable amount of effort.
- Increased profitability. This can be achieved through a combination of cost reduction and revenue enhancement.

Many methods of process improvement focus on the last item. Their spotlight is on reducing costs to increase profitability. Reducing costs means often using fewer resources or cheaper resources. If you just zero in on this, you will fall into the trap of the methods. Emphasizing cost savings angers or frustrates employees, customers, and suppliers. The methods are doomed in advance to fail.

WHAT HAVE BEEN SOME METHODS OF PROCESS IMPROVEMENT?

It is possible to devote an entire book to reviewing all of the various methods of the past 80 years. Instead, some of the major methods will be addressed and commented on so that their experience in use can provide lessons learned.

- *Industrial engineering.* Beginning at the turn of the twentieth century in factories and later in offices, engineers devoted attention to the detailed manner in which employees performed detailed tasks. Work measurement methods were developed. There were noticeable results in the production line. However, in other situations, the results were insufficient in terms of the effort.
- *Total quality management (TQM).* Deming and others developed techniques for improving the quality of output and reducing the numbers of defects and errors. TQM contributed an understanding of the importance of quality. However, TQM results often have taken too long and not yielded large-scale results.
- *Continuous improvement.* Under this approach, you work on a constant basis to keep improving processes. This is a good idea and showed the importance of giving processes attention over time. However, it is very difficult to sustain anything that is continuous.

- *Computer systems and technology.* The approach here was to fix processes by automating transactions. This is at the heart and soul of much of the productivity gains of the past 50 years. However, people now realize that it does not give the entire answer. Many organizations have spent millions of dollars and not really gotten the promised benefits.
- *E-business.* If you think about it, e-business or e-commerce is the next step in IT. IT implemented online systems within companies. E-business moves the online systems and technology out to customers and suppliers. E-business is the next stage in automation. However, automating transactions from broken or crippled processes doesn't work well. One company automated its ordering process on the Web but did not fix the fulfillment process. The result—many dissatisfied customers.
- *Reengineering.* Ah, it sounded so good and so appealing. Just throw out the old process and begin with a blank tablet (shades of the philosopher John Locke). For many firms this was a disaster. The approach was seriously flawed. First, it did not take advantage of what was learned from the old process. Second, people change processes but not the underlying systems. Third, introducing too much change at one time disrupted the organization.
- *Specific systems efforts.* One example is enterprise resource planning (ERP). Here software firms claimed that their products could not only provide greater management information for planning and control, but they could also provide process improvement. Unfortunately, many of these systems did not even touch the processes. Instead, they created another layer of work to feed the ERP. One Asian firm got around this problem by dedicating two clerks to input data from existing processes into the ERP, thus minimizing the damage to the result of the company. Don't get us wrong. There are benefits to ERP. However, these have to be weighed with the planning for the processes and risk.
- *Downsizing.* The theory here is that if you make some of the people go away, you can then improve the processes. This sometimes works, but often does not. You just drive the good people out and may be left with gnomes, turkeys, or trolls.
- *Outsourcing.* If you can't fix it or don't want to, get rid of the process. This sounds appealing. However, outsourcing creates new problems. The skills and processes of the outsourcing firm can be questionable. Things may get worse, and since you don't control the outside firm anymore, outsourcing can create more management problems.

So there you have some methods. What are common threads here? First, the methods tend to focus on only one aspect of the elements of a business process. Thus, they are likely to run into trouble because they are not comprehensive. The lesson learned here is that you must pay attention to more of the elements of the process. Second, each has promised major benefits, but often the benefits are not easily replicated to other organizations. Keep in mind that surveys indicate that over 50% of reengineering efforts either only partially succeeded or failed. Similar statistics have been cited for other methods. The lesson learned here is that you must use commonsense methods that appeal to self-interest. An overall guideline is that you are most successful if you employ a balanced approach.

WHAT IS SIX SIGMA?

Six Sigma is one of the newer methods of process improvement. To understand the phrase, consider a bell-shaped curve that describes defects of produced services or products. Sigma is the Greek letter used in statistics for the standard deviation. Six sigmas refers to having 3.4 defects in a million activities. The phrase relates to the roots of the approach—total quality management and statistical analysis. Six Sigma aims at implementing processes that are: (1) comprehensive, (2) flexible, (3) sustaining, (4) understanding of customer needs, (5) measurable, and that (6) contribute to business profitability and success. No one can argue with this goal. The previously discussed methods worked to achieve similar results. So what is different about Six Sigma? One major difference is the focus on customer requirements and the customer point of view. Many of the past approaches to process improvement were internally focused, rather than externally centered. Each chapter in the first four parts of this book addresses how Six Sigma applies. Note that we will not be going into the details of all of the Six Sigma methods. We are focused on achieving the goals of Six Sigma that are excellent.

WHAT IS THE APPROACH FOR SIX SIGMA?

The formal approach to Six Sigma is based on collaboration, dedicated effort, ongoing involvement from upper management, measurement of quality, continuous improvement, customer focus, process redesign, and statistical methods. In formal application, upper management is introduced to the approach. Then work begins with extensive and intensive training in the Six Sigma methods. Training varies from three weeks or more (Black Belts) to two weeks (Green Belts). After the training ends, work begins on specific processes. A number of major corporations such as General Electric and Motorola have used Six Sigma with successful results. There have also been a number of failures.

However, there is an issue here. What if your organization does not have the resources and money to pursue Six Sigma? You can admit that the goals of Six Sigma fit with your objectives. Easy enough. But you cannot spare the resources. Taking critical people from their positions for intensive training and then involvement on a large-scale, in-process improvement can be a daunting thought. Some companies have tried radical change and devoted massive resources to process improvement methods and almost driven themselves into bankruptcy. Here is the dilemma. You want to achieve Six Sigma results, but cannot afford a vast investment. You also cannot risk failure. What do you do? That is the purpose of this book.

MYTHS ABOUT PROCESS IMPROVEMENT

From our experience and those of others, a number of myths have arisen over process improvement. Let's discuss some of these, indicate their impact, and explore how they can provide you with guidelines to apply to your own process improvement.

- *The myth of the magic method: There is a silver bullet method, we just have to find it.* Unfortunately, this is probably not true. However, people keep proposing such methods. Why? Sometimes because a method may have worked in several companies. Sometimes to make money. Sometimes to gain recognition. Keep an eye for new methods, but review them carefully. Answer the question, "What do the people who are proposing it to you get out of using the method?"
- *The myth of definition: If we can just define the process in great detail, then we can improve or automate it.* The real world is not that simple. Processes are almost organic. They grow, change, and transform themselves due to human beings. Processes are dynamic. This does not mean that you give up. You just have to recognize the situation.
- *The myth of the technical approach: You can employ technical or mathematical methods to do process improvement.* Well, they certainly are ingredients to process improvement. But have you ever seen a mathematical method deal with political factors? No. However, technical methods are important.
- *The myth of automation: You can resolve many process issues through automation.* For specific transactions such as those of automated teller machines (ATMs), this is true. However, it is difficult to consider that this can be done for most processes, given the state of the technology today.
- *The myth of management focus: If upper management gets process improvement going and sticks with it, there will be lasting improvements to the work.* This is only partially true. Management must be committed and somewhat involved, but there are just too many other things that managers have to do to spend too much time on improvement—even those managers with the best of intentions.
- *The myth of measurement: Detailed measurements can be employed to generate and maintain change.* This is fine for process control and other systems, but there are problems with processes that involve more manual effort. Measurement has the following shortcomings if carried out too much and in excessive detail: The method

of measurement and what is being measured can impact the work. People focus on what is being measured, such as volume or quality, and neglect other factors. Measurement may take too much time to carry out and analyze. Measurement has its place. However, it is not dominant.

15 REASONS WHY PROCESS IMPROVEMENT METHODS FAIL

It is useful to review some of the reasons why process improvement efforts fail. They can guide you on what to avoid so that you don't travel down the wrong road.

- *Jargon or buzzwords are used.* Many techniques (not just in process improvement) employ jargon and words that are alien to most employees. Here are some of the problems. In order to implement the method, you have to train many people in the jargon. This makes a lot of money for the trainers, but it takes people away from the work. Jargon leads to confusion. You have to rely on outside experts who know the jargon—like the Delphi Oracle of Ancient Greece. People have to keep explaining things in the jargon, which gets in the way of communications.
- *The focus is only on long-term improvements.* This is nice, but the only long-term certainty is that we will die. You have to deal with the short term as well. Moreover, if you concentrate on the long term, the situation can change. People lose interest because there are no results.
- *The focus is only on short-term improvements.* Everyone does this. You do it on the weekend around your house or apartment. You take shortcuts to save time and money. However, if you only deal with the short term, you do not generate lasting improvements. Often, the original problems come back—just like athlete's foot.
- *Attention is centered on only one process.* You have already seen the problems with this—you can damage other processes and make them worse.
- *All transactions in a process are considered.* If you do this, you will probably never finish. There are just too many exceptions and workarounds to deal with everything.
- *Automation is treated as the solution.* Just gather the requirements and build, buy, or modify software. That should do it. It won't and it doesn't.
- *The company becomes overinvolved in one method.* Because of the complexity of processes, you cannot rely on one method or technique.
- *This is too much reliance on consultants and outsiders.* You depend on doctors for treatment of medical problems. However, the doctor is fixing something specific. A process is more general. It will be there long after the

consultants are gone. You can use consultants, but like other resources you have to employ them selectively.

- *Too much attention is given to management support.* As has been stated, management has a major but not dominating role in improvement.
- *Too much effort is devoted to marketing improvement.* Some companies spend too much time trying to explain how good process improvement will be for them. Employees immediately become suspicious and start polishing their resumes.
- *Systems and technology are excluded.* Systems and technology are key ingredients to process improvement.
- *Too many measurements are conducted.* You can measure something to death. Your goal is to improve it as well as measure it before and after.
- *Concentration is placed on the process of improvement and not the business process.* Your goal is to refine a method. It is to get results.
- *There is too much reliance on fuzzy benefits.* When all is said and done, it is the tangible benefits that really count. If intangible benefits can be made tangible, that is fine.
- *Too much attention is centered on customers.* Do you really know the customers? Yes, you want to know quite a lot, but you can go overboard. Employees count too.

25 CRITICAL SUCCESS FACTORS FOR PROCESS IMPROVEMENT

Don't get depressed from the myths and reasons for failure. There are principles that do work and have worked for more than 2,000 years.

- Management's role is to kick off process improvement and then support it by dealing with issues and insisting on reasonable measurement.
- In doing process improvement, concentrate on both short-term gains (called here Quick Wins).
- Avoid jargon and use common sense.
- Appeal to people's self-interests. This includes your employees, customers, and suppliers.
- Gather experience and lessons learned as you do process improvement to aid you in getting results.
- Manage overall process improvement as a program that is ongoing. Manage specific process improvement as a project.
- Be aware of and take advantage of the politics of the organization.
- Use systems and technology to the extent that is practical and feasible.
- Consider processes in groups, not individual processes.

- Conduct reasonable measurements at the start, during, and after process improvement.
- Implement a regular method of measuring key processes.
- Involve as many employees as possible in process improvement. This will gain more widespread support.
- Employ consultants on a targeted basis as agents for change.
- Give attention to steering the culture toward change and improvement through the process improvement projects and programs. This will impact how you organize the effort.
- Don't consider all transactions and work in the business processes. Concentrate on mainstream work and some frequent and critical exceptions.
- Pay attention to changing policies. Change here can give you Quick Wins.
- Avoid pulling people away from their normal work assignments to be involved in the process on a full-time basis for an extended time. This can weaken the process and isolate the people from the work. It can also make you overreliant on specific people.
- Concentrate on the knowledge transfer among employees to help junior employees.
- Involve junior employees more in process improvement.
- Concentrate on obtaining grassroots support for the new process. This is key to sustained results.
- Design new processes and work with maintenance in mind.
- Implement major change in waves. People can only take so much change. Start with the Quick Wins. Successive waves of these can lead smoothly to long-term process improvement.
- Establish management support among multiple managers. When you rely on one person, the effort may go up in smoke when the person leaves.
- Keep a sense of humor. Work can be fun and enjoyable. Process improvement can raise morale.
- Have several leaders involved in the improvement effort. Relying on one person may burn out that person and leave you with no backup. Moreover, due to politics it is useful to have several leaders—one of which is in charge at a given time for accountability.

There will be more success factors as you read further. However, these will get you started in thinking about process improvement.

THE GOALS OF THE APPROACH

These success factors and earlier discussion point to some very specific goals for your process improvement effort:

- Employees at the bottom as well as their supervisors support the integrity of the process, regardless of management. They see this in their own self-interest.
- Process improvement fits in with the culture of the organization.
- People see the importance of processes and are more aware of process problems and issues. They do not see process improvement as a threat, but instead as an opportunity.
- Collaboration and sharing of knowledge are more prevalent during and after process improvement.
- Employees make suggestions for further improvement openly. They see that better processes make their working lives better.
- The political environment of the organization is supportive of improving processes.

Experience has shown that if these goals are attained along with the critical success factors, you will get the tangible benefits listed earlier.

Here are some things we assume about you, our reader:

- *Resources.* You have limited resources. People have to do most of their normal work. There are few spare resources.
- *Time.* You have limited time. Time in process improvement is your worst enemy—more than money.
- *Politics.* There are political factors within your organization. Resistance to change lurks under many rocks.
- *Money.* You have limited money. You cannot afford the big study that does not pay off.
- *Priorities.* There are competing priorities for resources and direction.
- *Results.* There is pressure for results. This is not only from managers; it is also from employees.
- *Management support.* You do not have unlimited management support. Management has many other things to do.
- *Resistance.* Some processes on the surface appear fine and don't seem to need improvement. Another way to state this is that many feel "We have been doing the work the same way for years so why change?"

If these things sound familiar, you are in the right place.

A DOWN-TO-EARTH APPROACH TO PROCESS IMPROVEMENT

This book is not about just how to plan process improvement—it is about planning, doing, maintaining, and expanding what you improve. It does little good to improve something if it returns to the way it was after you leave. There is no silver bullet.

You logically start with understanding your business. This not only means reviewing the vision, objectives, and mission of the company; it also means identifying which processes are important and most in need of improvement. You also have to gather information on IT and technology as well as the industry and competition. There are political goals here. You are getting people to recognize the need for change. You are also making them aware of the outside world, which is valuable since many managers and employees are internally focused. This early work must be done in weeks, not months, and it can be ongoing.

You are next faced with selecting specific processes, work, and transactions in the business. Messing up here can be fatal later. You examine the work in more detail. Be careful. It is too easy to get carried away and gather too much information. Remember that the more that you collect, the more people expect. Here you will measure the current processes. You will develop a process plan. The process plan tells people what the future could look like some day. This is not what is easily attainable. It is what you are working toward. You will define the new business process and short-term improvements to the process (Quick Wins). As before, there are political objectives. By defining the processes, you are getting people to focus. By delineating the long-term future, you are showing the vision of the future. Examining the current work exposes the warts, problems, and issues—making people more eager for change. By developing the new process, you are getting people aligned with change and showing how the goals of the process plan can be supported.

Here is where many books and methods end—on the brink of implementation. But that is where the real enjoyment of process improvement begins. Some methods would now recommend that you plunge in and make changes. Wait! You have to develop an implementation strategy. You have to have a plan for organizing the effort to achieve cultural change and deal with the politics within your organization. Then you can develop the implementation plan. These steps achieve useful goals politically. First, they show people that you are serious. It is not just a study. Second, carried out right, you will ferret out points of resistance to change that you will have to address during implementation.

Implementation of process improvement involves two interrelated parts. First, you will be putting in place Quick Wins. These are short-term changes that yield tangible benefits. Second, Quick Wins are consistent with long-term process improvement. You implement several waves of Quick Wins in most cases that lead to the improved process. The political goal here is to gain support for the change and enthusiasm for the processes. Hopefully, you will make the employees fans of the processes. Note that we did not say fans of process improvement. That is asking too much. People seek stability. Measuring the results of the work gives you the political support for maintenance and ongoing efforts.

An area that is frequently overlooked is process maintenance and enhancement. As you have seen, processes deteriorate over time without attention. You will see specific techniques for maintaining processes and expanding improvement. Remember that people get worn out from doing this, so you have to organize periods of calm between the changes.

Let's go beyond the technical approach to the details. You want to concentrate on developing analysis tables, lists, and charts during your improvement work. The analysis tables will be called improvement tables so that they stand out. These things can be easily created, updated, and presented—better than text. You will be defining a series of comparative improvement tables. Each chapter will add more factors to consider—generating more improvement tables. These tables and charts are very important for helping you achieve political objectives in the work. You will also note that the book is written in a simple business style. That should be your style—simple and direct.

Figure 1.1 gives an overall view of the approach. While the diagram shows things moving sequentially, this is just the overall flow. In reality you tend to do more in parallel, giving the limited time and budget that you have. Notice that the steps and items are not numbered. This is by intent since we do not want to convey or reinforce the impression that these steps must be taken sequentially.

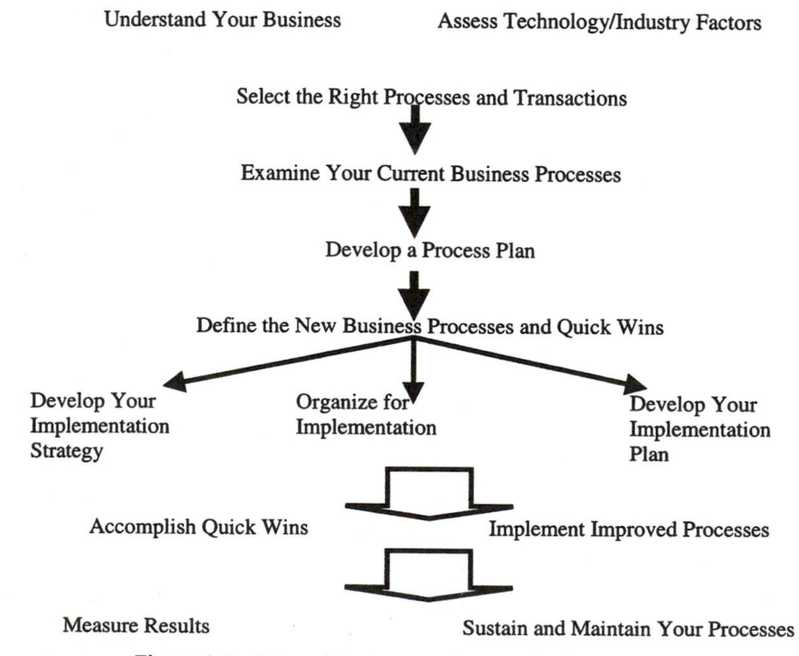

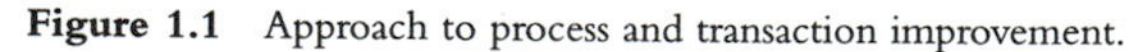
Figure 1.1 Approach to process and transaction improvement.

Finally, the approach is flexible to support two management alternatives to change and process improvement.

Strategy 1: General Process Improvement

Following this strategy, an organization seeks to identify potential improvement opportunities across the organization. No process is left out of consideration. This is the approach that is often pursued if the trigger or forcing factor for process improvement is internal management. An example occurs in a company where things are okay, but there is a feeling that improvements are possible.

Strategy 2: Directed Process Improvement

This strategy targets specific areas of the business. The strategy is often employed in reactive situations where there are demands for cost cutting and downsizing due to external pressure. Alternatively, there may be problems in a specific business unit.

WHAT IS DIFFERENT ABOUT THE APPROACH?

If you compare the previous approach with existing methods for improvement, you can find several differences.

- *Politics.* The approach specifically addresses political issues and the politics of change. Our experience over the past 30 years consistently reveals the importance of political factors and the political dimension of change.
- *Quick Wins and change.* Focus on Quick Wins as well as long-term improvement. Management and employees often have this in common—they don't want the change to stretch out in time. They want to get on with their work.
- *Incorporation of systems and technology.* You often cannot get lasting change and improvement without implementing improved systems.
- *Going beyond a customer or specific focus.* Our approach considers customers, employees, management, the processes, the external environment, stakeholders, suppliers, and other perspectives.
- *Attention to transactions.* Processes are really too general. The work is performed in transactions. So when you get down to it, the "devil is in the detail." That is why we center the approach on transactions. In many cases, you cannot afford and do not want to change an entire process. You may want to leave some exceptions and workarounds to die on the vine.

• *Collaboration.* Every method involves employees and managers. The difference here is the extent of involvement. Success in both the short and long term can only occur if the people doing the work are committed to the project. This does not mean full-time involvement, but meaningful and directed involvement. Working together raises morale and increases motivation.

• *Table-and-chart-based method.* The method applies common sense. There is no jargon—only standard business terminology. Rather than generating massive documents that no one will read, the attention is on developing charts, tables, and lists that you can employ throughout your improvement effort.

• *Limited resources.* We don't assume that you have infinite resources and time. We don't assume that you are part of a huge corporation. You could be a small business or department. We do assume that your personnel resources are limited.

THE ORGANIZATION OF THE BOOK

The first four parts of the book are organized consistent with the approach. Each chapter has the following sections:

- *Introduction.* How Six Sigma fits with the topic is covered along with providing background.
- *Objectives.* Each part of process improvement involves technical objectives, business objectives, and political objectives. Remember, you just can't ignore the political realities.
- *End Products.* This is a list of what results you want to get out of the work in a particular chapter.
- *Methods.* Here is the detailed approach for the subject. Specific techniques are given for technically doing the work along with how to achieve business and political objectives. In fact, this section always ends by addressing a political goal relevant to the chapter.
- *Examples.* The four examples summarized in the next section are followed and discussed throughout.
- *Lessons Learned.* These are specific guidelines for carrying out the methods. These tips have been gathered from real-world experience and should help you to overcome hurdles.
- *Problems You Might Encounter.* From our experience we list detailed situations and problems and what to do about them.
- *What to Do Next.* These are specific steps that you take after you read the chapter.

The last part of the book addresses specific issues that you are likely to encounter in your process improvement effort in general. Here you will be examining management, business and organization, work, and political and cultural issues. Additional specific issues are covered in the earlier chapters. For each issue in Part V, you will be considering the following factors:

- The impact of the issue if it is not addressed. What damage is possible?
- Prevention of the problem or issue. How can you head off the issue before it hits you?
- Detection of the issue. How can you know when the issue is looming?
- Action on the issue. What realistically can you do about it?

EXAMPLES

The four major examples are presented next. Each is a composite from experience with several firms in that industry. In addition to these, there will be examples from insurance, agriculture, and energy. Other industries will also be discussed. In these examples we will examine the politics as well as processes. You may initially think that these companies are sick. From our experience with more than 140 firms and agencies, they are typical. Most books examine cases or examples with an academic view. Here you will dig deeper into the politics and real-world environment. Keep in mind Bette Davis's classic statement in a movie that we paraphrase here—"Fasten your seat belt, it's going to be a bumpy ride." You will peer into the political world in each company. If you feel that these examples strike too close to home, relax! Many people are in the same boat—regardless of the country.

ASC Manufacturing

ASC makes specific large-scale engineering structures, such as aircraft, large pieces of equipment, and satellites. ASC has tried a number of methods to make improvements, but they did not work. People have been burned before. There is a reluctance to try something else. But ASC must improve its competitive position or die on the vine, so to speak.

Kosal Bank

Kosal Bank is a large full-service bank that provides credit card, installment lending, leasing, and other services. Each area of the bank has grown its own procedures and processes. There are many inefficiencies. Customer satisfaction is low.

Hetsun Retailing

Hetsun operates retail stores and a catalog service. The company wants to go into e-business and at the same time wants to improve its processes. Like ASC, Hetsun has tried several methods, but they only partially worked. None of the methods produced lasting results.

Lansing County

Here our focus will be on county operations. Much money has been spent on automation with little change in processes. Management is frustrated. Experience reveals that Lansing is bureaucratic, but not more than other agencies or large companies. There are many political impediments to change.

Lessons Learned

Lessons learned are specific tips related to the chapter. Here are several from this chapter:

- Keep a low profile on process improvement. The more attention you give it, the more people expect and the more they feel threatened.
- Dampen expectations for fast results by emphasizing the issues in the processes. Responsibility for results rests with the employees and supervisors as well as with management and the process improvement team.
- Politically orchestrate when the improvement work will be visible. Remember that the project leader is the director of the movie and affects timing.

WHAT TO DO NEXT

This section identifies opportunities that you can explore after reading the chapter. Answer the following questions about your organization:

- What previous efforts have been made at improvement? What happened to these? See if you can determine what was behind the outcome.
- What measurements have been made for the processes?
- Consider a specific process that you are familiar with. What is the extent of exceptions, workarounds, and shadow systems?
- What process training exists for employees?

CHAPTER 2

Understand Your Business

INTRODUCTION

It is tempting to plunge in and identify the processes to improve as well as implement. Many books and approaches do just that. They assume that management support is all that is needed to get going. This is one of the reasons why many improvement methods and efforts fail. The situation resembles that of an alcohol addict. The person must first admit to having a problem before he or she can begin treatment. Here you want to not only understand the dynamics of the business, but also you want to find out what is bothering people. If your improvement effort addresses some of what is causing people grief, then you will get more support.

Chapter 1 emphasized that political support is essential for success in improving processes. Let's begin this chapter with a better understanding of a business process. Most business processes are composed of transactions. A transaction is a specific set of steps to carry out the process under defined conditions. Most, but not all, processes are described and characterized by their transactions. A process is interdependent with the organization departments, other processes, and computer systems. Consider Figure 2.1. In this diagram you can see the interdependence of processes among each other. You can see how automation and the organization support and carry out the transactions. Management is shown at the top of the diagram on purpose. Most managers realize that their future is heavily dependent on the effectiveness and efficiency of the processes and transactions.

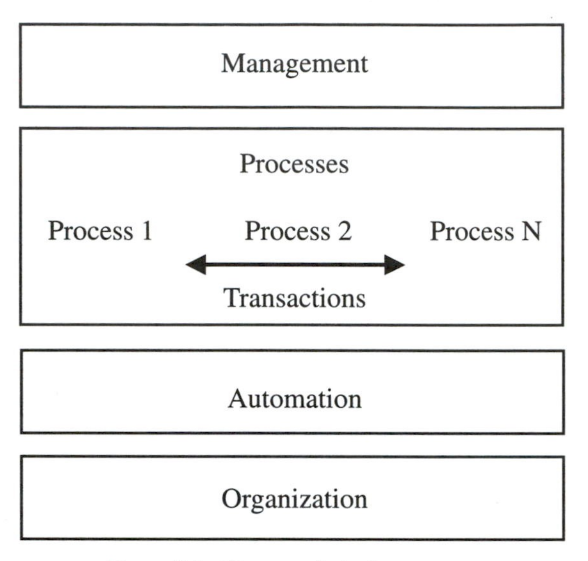

Figure 2.1 Diagram of a business process.

The figure highlights some basic points:

- Process change must be carried out with the support of not only management, but also departments and the information technology (IT) organization.
- Changing a process means carrying out transaction changes at the nitty-gritty level.
- Because people's lives and power are tied up in processes, process change tends to be very political.

One problem continually arises in process improvement and other areas such as IT—how to relate general business and other goals and strategies to very specific process, business, and technical issues, actions, and plans. Figure 2.2 illustrates how this book will address this issue. At the center are the business processes and transactions. Eight spokes lead to the outer part of the diagram, which comprises business, organization, and technical areas. The key idea here is to relate business, organization, IT, and other factors to each other through the business processes. This is in line with the previous chapter, which presented the role of processes and transactions. How will you use this diagram?

- You can relate business goals, strategies, and issues to processes.
- You can relate organizational issues, structure, and management factors to business factors at both the enterprise and department/division level through processes and transactions.

- You can relate IT and other support activities to the business through the processes and transactions.

What is the benefit of this? When you improve processes, implement e-business, or carry out other improvements, these are related to the day-to-day business activities such as processes and transactions. A basic question is how these actions help support business and other goals. The diagram provides the mapping. Next, let's back up and select which activities to improve. As you will see, there are a number of alternative approaches. Which to choose? What is the best way? Answers to these questions require the same mapping as that shown in Figure 2.2. Working with the figure in general is fine for discussion purposes, but how do you get this down to the nitty-gritty? You do this through the improvement tables. You will be creating tables of relationships to link the various parts of the approach together. You will be creating a web of relationships that has a number of benefits for you:

- Employees, vendors, and customers can better understand what is going on and be more supportive and involved if they can see the big picture and how details relate to general goals.
- There is more support for changing specific transactions in general if their importance can be understood in terms of the departments and enterprise.
- IT changes and problems can be better understood if their impacts on the business and processes are known.

Thus, drawing up and using the improvement tables helps you to gain understanding, support, and commitment. In this chapter, you will work on understanding critical business factors at the enterprise and business-unit levels. The questions you will raise are "What do you do with the information? How does it contribute to process improvement?" To answer these questions, you will create lists and improvement tables that you can then employ in later stages of the work.

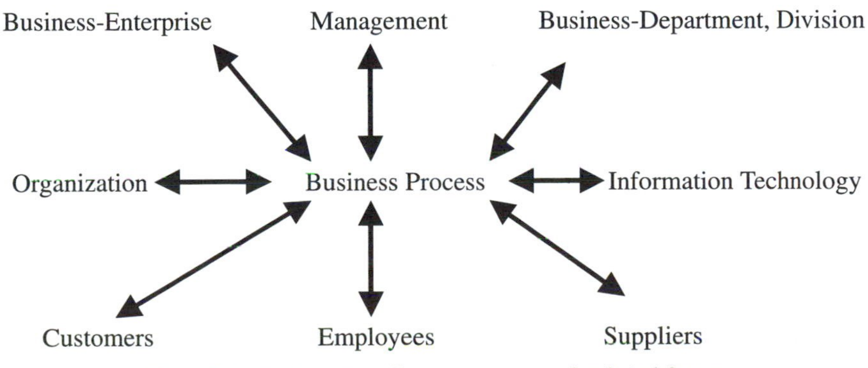

Figure 2.2 Relationship of business, process, and technical factors.

OBJECTIVES

Technical Objectives

Most people tend to focus on understanding the business by interviewing higher-level managers and reviewing the annual plan, business plans, and other documents. Using this academic approach, we would then define what the important goals and issues are, which would later help us to select specific processes and transactions for improvement. It sounds so good in theory. Then you get out there to do it and all hell breaks loose. People ask you, "Why are you asking these questions? You are supposed to be improving processes." Moreover, you find disconnects between the goal and the reality.

So let's get practical. Your technical objective is to make lists of objectives, strategies, and issues at the enterprise and business-unit levels. You will achieve technical success if you have a fairly complete list that you have validated. Notice the word "fairly." We use this word because our many process improvement efforts have taught us that you never have a totally complete set. Moreover, just compiling the lists and creating improvement tables may result in the organization having to *define its objectives, goals, and strategies with more precision*. We emphasize this point because it is a major benefit of the work. In this book, we want to give you as many side and additional benefits as possible. The more benefits you deliver, the more support you will get.

Business Objectives

What do you hope to get out this effort from a business view? First, you want to immerse yourself in the business overall and at business-unit levels. Even if you have worked for a company for 20 years, you may tend to miss the big picture since all of us work at the detailed level.

A second business objective is to prepare for the detailed analysis that will come. If you plunge into the detail too soon, you can become lost in this detail. Then you implement what seem to you very worthwhile improvements, but they do not help the company overall in terms of achieving their goals. In other words, you healed part of the patient, but the patient is still on life support.

Political Objectives

As you will see in each chapter, there are significant political objectives. Here are some of the critical ones:

- Get people to understand why process improvement is needed.
- Have more managers and employees involved in the analysis from the beginning.
- Validate and get people to agree to the mission, objectives, and other aspects of the enterprise.

Let's look in more detail at each of these objectives. Take the first one. Didn't management approve process improvement efforts? Why do you need to have people get behind the improvement effort. Because most people are comfortable with what they do. They don't see the need for change since they perceive at their level (and rightly so) that their processes are working. After all, people are getting paid, shipments are going out, sales are being made. That is why management has to—*from the start*—get individuals to understand and support the need for and benefit of improvement *at all levels.* This is a theme of the book from our experience—the need to do constant marketing of process improvement. Process improvement does not sell itself—even after the benefits and methods have been defined and adopted.

The second objective leads to scope of involvement. From the beginning you want to get middle- and lower-level employees involved. In e-business you might involve in addition selected suppliers and customers. Why is this necessary if you have support from the top? Because top-level support is not sufficient when you get down to the detailed changes. At the detailed level, a single clerk in a strong informal power position can block improvements.

Some might also question the third objective. The objectives and other factors have been defined and enunciated to employees for many months or years. Why do we need agreement? Because these factors may not be complete. They may have been lip service, but may not have been followed.

END PRODUCTS

What results do you want from this work? Technically, you want to develop the following:

- Specific lists of business mission, vision, objectives, and issues
- List of critical business processes
- Tables relating these items to each other
- An organization score card that represents an internal measurement of the firm prior to making process improvement changes

On a wider level, you want to gain an understanding of what has been done in the past. What worked and what did not work? You also want to gain concurrence from managers and employees about the issues and problems that they will face. The organization score card is a means of assessing in tangible form where you are prior to implementing change. This is essential for both proactive and reactive reasons. Proactively, you want people to always be aware of what the past was like. Remember the old saying that "if you don't understand the past, you are doomed to repeat it in the future."

METHODS

THE DRIVERS FOR IMPROVEMENT

You must begin with the factors that created the process improvement effort. These can be a combination of external and internal factors. Here are some examples:

- There is external pressure from investors, bankers, and others to cut costs.
- New management is at the helm and wants to make a quick impact on results.
- Competitors are doing much better than your company is in terms of sales and profitability.
- There is substantial employee turnover. Morale is low.
- Sales have increased, but costs have risen even faster.
- A lot of money was spent on some initiative (enterprise resource planning [ERP], e-business, total quality management [TQM], etc.), and results were not as great as expected.

Use this list as a checklist to examine why the process improvement effort has been initiated.

Does it matter if the drivers are internal or external? Yes. External drivers tend to be more pressing. There is a greater sense of urgency. Also, external drivers tend to be trailing indicators. That is, the company is already in trouble and it is not evident on the outside. Here are some basic truths:

- Ideally, you want to do process improvement due to internal drivers, since these tend to give you more time for improvement. Also, internal drivers indicate that management and employees see the need for change.
- External drivers mean that the company managers may have thought that their processes were just fine and that no change was needed.
- Making process changes due to external factors tends to be reactive; making changes due to internal factors tends to be proactive.

Now we come to a basic point. Many firms focus on short-term profitability to keep their stock prices up and maintain investor happiness. These are standard goals. Often they translate into "cut costs," which in turn means layoffs and terminations. Firms set targets for savings. Next they apportion the cost cutting across business units. Then the head cutting sets in. While this approach has been going on for centuries and is a standard one, it tends to be counterproductive for a number of reasons:

- The good people tend to leave. The group that is left consists of employees who cannot readily find other work. To put it another way, you may lose the Einsteins and be left with gnomes and trolls. Not good.
- With this approach, no attention is paid to the processes so that cuts are made that tear apart critical processes. Some companies never recover.

Examples of this approach can be found in many failed dot-coms where indiscriminate cuts were made. The companies then later folded their tents and disappeared in bankruptcy.

So what is a better approach? Make the cuts based on business processes. Identify which processes can be reduced or simplified and then initiate the cuts based on process and transaction analysis. But some will say, "We don't have time to do this." The response is that it does not take that much more time.

Why do companies get into trouble? Because they take their processes for granted. They stop making improvements because times are good and orders are flowing in. This is a natural human trait but one that has to be resisted. To see how old this tendency is, consider ancient Rome. Once Rome was on top of the world, at the end of the first century A.D., innovation began to stop. Major reasons for the decline of the Roman and Egyptian empires were that innovation and change dried up. The status quo was reinforced.

Gain Management Support

Many believe that getting top management support at the start of any initiative is enough. It is not. Top management support can be fleeting, as these managers go on to other issues and problems. You need to build management support for change all along the way. How do you do this at the start? One way is to show the potential for technology and new systems. However, technology has often promised much and delivered little. Here is a fundamental thought.

You tend to build more support for process change by showing and getting acknowledgement for the ways things are today than by presenting the fuzzy promises of the future.

Consider drug and alcohol addiction. Telling addicted people how good they will feel if they kick their bad habits sounds good, but they still have the addictions. They must understand that they need to change. It is a basic tenet of rehabilitation that people acknowledge that they have problems and the impacts of the problems.

Where to Start

Enough with some general thoughts; let's get down to work. You don't want to rush out and start interviewing and collecting data. Much data and information are already available. Let's make a list of what you should do:

- Plan your process improvement effort.
- Review existing information such as plans, documents, previous studies, and so forth.
- Evaluate previous efforts.
- Set up the tables for process improvement.
- Review the mission and vision of the organization.
- Assess business objectives at the enterprise and major business-unit levels.
- Identify key business issues.
- Collect information on key business processes.
- Build the organization scorecard.

These are not sequential activities. You want to try to accomplish these in parallel. Doing the items on this list will support getting the end products defined earlier.

Therefore, you will start by vacuuming up information—even while you are planning your process improvement work. Here is a list to get you started:

- Last three annual reports
- Last eight quarterly reports
- Reports by major business units on their activities and results over the past one to two years
- IT plans and reports
- Memoranda, e-mail, and other documentation on problems and issues.
- Web-available information on the company

Go to your favorite search site on the Web, such as Yahoo or Excite, and conduct a search on your company. When you find the financial results, you will also find comments. These are often called "insider comments." They are remarks by people who are happy or unhappy with the company. Print these out or gather them into a file. Keep in mind that these comments can come from disgruntled employees and others who have a personal ax to grind.

PLAN YOUR PROCESS IMPROVEMENT WORK

You need to have an overall process improvement plan. This means that you need to have a project plan, databases for issues and lessons learned, ways to structure and organize information, and software tools to help you do the work. Since process improvement is a collaborative effort, you will want to define how people will work together. Gather the following information:

- A project plan for the improvement effort.
- A description of how the effort will work.
- A database for issues and problems. Issues are things that you have to address as part of process improvement.
- A database of opportunities and Quick Wins. This is where you capture ideas for improvement.
- A database of experience and lessons learned.
- Software tools that will be used for supporting the work.
- Organized files.
- Outlines of major deliverable items that affect the work.
- A plan for how you will report to management.

The need for these items is evident. You want to define these elements at the start for several reasons. First, you want to be organized from the beginning. Second, by showing you are organized and that you have structure, you will get more political support.

Here are some basic guidelines for project management. You can find more in the book by B. P. Lientz and K. P. Rea, *Project Management for the 21st Century*, Third Edition (San Diego, CA, Academic Press, 2001).

- Create the project plan with the involvement of the team members who are participating. This will not only gain their commitment, but it will also give you a more accurate picture of the challenges ahead.
- Keep the level of detail in the plan reasonable. Each task should take one to two weeks to execute.
- Have the team update the plan twice a week.
- Associate issues and problems with tasks in the plan. Do the same for experience and lessons learned. This will give you a more complete and integrated plan.
- Set a baseline plan at the start and measure your progress against the baseline.

"How the effort works" is a vague expression that deserves some attention. It is here that you explain the roles and responsibilities of people and organizations in process improvement. You will define how people are to work together. You will delineate the method for process improvement. In short this is where you explain how things are to work (Figure 2.3).

Area	Task Group	Milestone
Planning	Approach	Description of method
	Roles	Roles agreed to
	Plan	Plan completed
	Reporting method	Implemented
Understand the business	Mission and vision	Analyzed for tables
	Objectives	Analyzed for tables
	Issues	Identified and agreed to
	Key processes	Identified
Technology and industry	Current technology	Issues identified; IT score
assessment		card
	Potential technology	Opportunities determined
	Competitive assessment	Key competitive factors
		determined
	Industry trends	Industry score card
Process selection	Determine process groups	Complete list of groups
	Selection of processes	Agreement on processes
	Selection of transactions	Agreement on transactions
Process evaluation	Detailed analysis	Process score cards
Process plans	Develop process plans	Process plans
	for future of processes	
New processes, Quick Wins	Determine short and	Agreement on changes
	intermediate term change	and priorities
Implementation strategy	Develop staging method	Strategy approved
	for change	
Organize the	Define the detailed	Gain support and
implementation	approach and method for	involvement
	change	
Implementation project	Develop the detailed plan	Completed and approved
plan		plan
Quick Wins	Implement Quick Wins	Measure results
Process change and	Implement process	Measure results
improvement	change	
Measurement	Measure results	General measurement

Figure 2.3 Major task areas and milestones for process improvement.

Figures 2.4 through 2.6 give the data elements for the databases of issues and problems, opportunities, and lessons learned, respectively. You can use a standard database management system or spreadsheet to set these up. You can use the issues in Chapters 13 through 16 as a start for the issues database.

Software tools will determine how you will do calendaring and how you

- Issue identifier
- Title
- Type
- Status
- Date created
- Description
- Impact if issue not addressed
- Area affected by issue
- Decision taken on issue
- Actions taken
- Date resolved
- How resolved
- Comments

Figure 2.4 Data elements for issues and problems.

will communicate through memos, e-mail, voice mail, and analysis. Figure 2.7 presents a sample file organization.

We don't want you to spend all of your time documenting, but you do need to define the contents of the major deliverable reports and presentations that you will give. We have listed these items in Figure 2.8. Of course, the list depends on your specific situation and scope. If, for example, you are going to make major systems changes, then there would be more task areas and milestones for IT-related items on your list.

- Opportunity identifier
- Title
- Description
- Type
- Date created
- Who created it
- Area to which it applies
- Status
- Expected results
- What happens if the opportunity is not pursued?
- Decisions taken
- Date of decision
- Actions taken
- Date of actions
- Results achieved
- Comments

Figure 2.5 Data elements for opportunities and Quick Wins.

- Identifier number
- Title
- Area or type
- Status
- Date created
- Who created it
- Description
- To what it applies
- Suggestions
- Expected results
- Benefits
- Guidelines on how to do it
- Comments

Figure 2.6 Data elements for lessons learned.

In management reporting, you want to include the following:

- A summary project schedule
- Budget versus actual analysis
- Accomplishments in the previous period
- Upcoming accomplishments
- Outstanding major issues

Note the last item. You want to keep management involved of the problems and issues that you are facing in the work.

Establish the Process Improvement Organization

You can set up the organization in any way you want. If you want to follow the approach presented in the previous chapter, you will need to identify three levels in the process improvement organization:

- *Executive-level committee.* This is composed of high-level managers who assemble when needed to resolve major issues, make or endorse decisions, and approve the direction and results of process improvement.

- Initial project plan for process improvement
- Process improvement kick-off documents
- Issues related to the process improvement work
- Memoranda, e-mails related to the work
- Budget reports
- Updated project plans and progress reports

Figure 2.7 Sample file organization.

- Management
  - Process improvement approach
  - Process improvement organization
  - Process improvement plan
  - Process improvement progress reports
  - Method of selecting processes and transactions
- Technical and business
  - Organization score card
  - Tables related to business factors
  - IT score card
  - Industry score card
  - Technology and industry tables
  - Process group candidates
  - Processes selected
  - Transactions selected
  - Process score cards and evaluation
  - Process plan
  - New transactions and processes
  - Exceptions, workarounds, and shadow systems and actions
  - Quick Wins
  - Implementation strategy
  - Implementation plan
  - Measurement of Quick Wins
  - Measurement of new processes and transactions

Figure 2.8 Major deliverable items for process improvement.

- *Steering committee.* This is a committee of managers, supervisors, and employees that serves to oversee the improvement effort.
- *Action teams.* Each team is composed of four to five employees who focus on a specific area such as marketing, sales, finance, or IT.

In addition, you will have a core process improvement project team composed of the project leaders and team members who will be coordinating the effort. This corresponds to the concept of the project office in project management.

The structure is shown in Figure 2.9. This three-tiered approach has a number of benefits that have resulted time and again from our past and present improvement efforts. They include the following:

- Having the steering committee provides a link between management and the employees, as well as a political buffer.
- The steering committee can address tactical concerns, while the executive-level committee can address strategy.
- Three levels provide a means of addressing escalating issues and opportunities in an orderly manner.

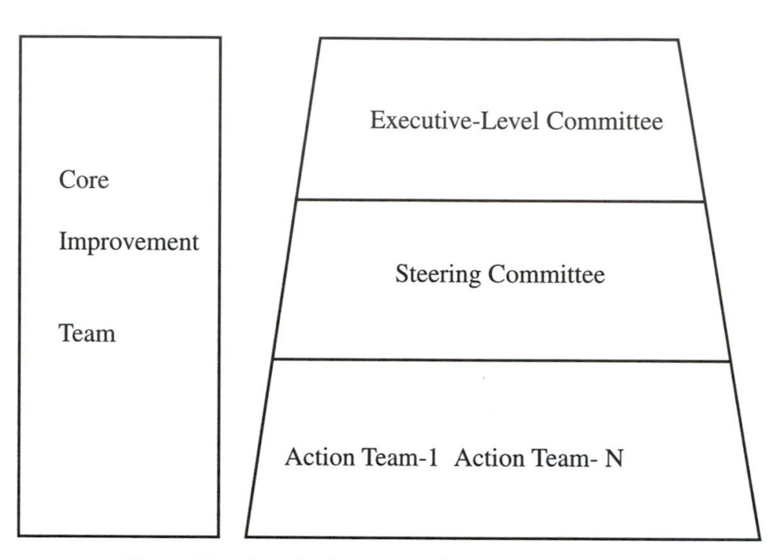

Figure 2.9 Organization structure for process improvement.

- The three-tiered approach is better than a two-tiered one because it provides buffers between levels and it does not involve upper management in excessive detail.
- The approach tends to involve more people. This means more coordination, but it also means that you will get more grassroots support and participation.

Some guidelines and explanation for the structure are needed:

- Action team members do their regular jobs and duties as well as serving on the team. This puts pressure on them to get things done. They also do not see their jobs as being at risk.
- Each action team is composed of four to five people drawn from different departments.
- The leader of an action team comes from an area that has a different focus from that of the action team. For example, the head of the sales action team might come from production or finance, but not from sales. This provides an impetus for change.
- Each action team has a subject matter expert who is familiar with the workings of the specific department or area.
- Whenever possible, each person serves on only one action team. This means that you will involve more people in the work and get more support.
- There should be common members between the executive committee

and the steering committee and between the action teams and the steering committee. This provides continuity.

It is a basic premise of process improvement from our experience that the way you set up the process improvement effort is critical to the degree of change and benefit you end up with. Note that this is true with the action teams. Each action team is headed up by someone who does not have a vested interest in that area. In fact, it might be someone who has concerns or complaints about that area.

How does this organization work and function? The steps are as follows:

- The improvement project team oversees and coordinates the work. It also does a substantial part of the analysis. Team members might, for example, do preliminary analyses of specific change opportunities to serve as models for the action team.
- The action teams work to select opportunities for process improvement. This can be done in two ways. They can identify opportunities at the beginning before the processes are selected. Alternatively, they can work on opportunities after the processes and transactions have been selected. The method depends on whether the overall approach to process improvement is targeted or general.
- Once opportunities are identified, they are written up in terms of the general situation, the impact if not addressed, the benefits, the implementation approach, and the barriers to implementation.
- The action teams now vote on which opportunities they will focus on in their own work.
- The project team then organizes these results and passes them to the steering committee, where another vote takes place. The project teamwork ensures that there are going to be cumulative benefits and not just disjoint changes.
- Having decided on the most promising improvements, the action teams now prepare more detailed business cases for each opportunity. This includes more description of how to implement the improvements and quantify the benefits.
- The business cases are reviewed by the project team and finance department to assess the benefits.
- The steering committee now votes on the business cases. Any revisions are made.
- The executive-level committee now votes and reviews the business cases and approves implementation.
- The project team now organizes the implementation.

This method ensures participation by employees and management in a structured way. Extra work is avoided by not considering some of the opportunities either because of lack of benefits, lack of contribution to objectives or issue resolution, or difficulty in implementation.

Evaluate Previous Efforts

With the approach outlined, we can now discuss how to do the initial work. You have gathered information on previous efforts at change. These include implementing systems, technology, quality programs, and other changes. In some Six Sigma efforts, failure occurs because people proceed down the same path as they took in previous attempts. The result is the same—effectively, nothing changes. For each effort you want to answer the following questions:

- What was the original goal of the effort or initiative?
- What were the actual results?
- Were measurements taken before and after the initiative?
- Why did it fail?
- Why did it succeed?
- What additional impacts were felt?
- Who were the key players in the initiative?
- Looking back, what could have been done differently?

Obviously, you cannot answer these questions yourself. You will have to do some informal interviewing to gather more information. We stress informal because if you make formal interviews, people tend to be defensive. They may think that you are conducting a witch hunt. You should give as the reasons for gathering the information that (1) you don't want to repeat failed efforts and (2) you want to reuse experience and methods that proved successful.

The results of this work will be the following:

- Identification of people who are supportive or resistant to change—very useful later.
- Lessons learned on the organization's attitude toward change and measurement of change as well as management support of the efforts.
- Hints and clues as to what areas might be good targets for process improvement.

Review the Mission and Vision

What do the words "mission" and "vision" mean? You want to answer this question because the terms are often very fuzzy. The vision is where you want to end up after you carry out the mission. So if your mission is to make customers happier, then the vision is for satisfied customers who do repeat business with your organization.

To avoid dealing with fuzzy generalities, let's get down to basics. Any mission or vision statement should say something about each of the following:

- Stakeholders—shareholders, investors
- Management
- Organization
- Employees
- Product or service
- IT
- Customers
- Suppliers
- Processes

When you analyze the mission and vision, you should create an outline under these bulleted items. Under each item, you write "the intended result" and "benefits." The first is from the mission; the second is from the vision. Alternatively, you can create a table like the following:

Area	Mission	Vision
Stakeholders		
Management		
Organization		
Product/service		
Employees		
IT		
Customers		
Suppliers		
Processes		

What does this table do for you? Eventually, you will be selecting processes and transactions. It will be useful for you to determine whether making changes there will have an impact on the vision and mission.

When you have created the table, it will be useful to have these points reviewed. The review will be an eye opener—trust us. When people see gaps and holes in the interpretation of common terms, there will be a reaction. That is okay. What you are doing in this early stage is supporting a dialogue and communications. In several of our improvement efforts, this yielded tangible benefits by improving the mission and vision.

Assess the Business Objectives

From annual reports and other sources, you can make a list of business objectives. These tend to be fuzzy, like the mission and vision. You need a way to organize these objectives. List them in the following categories:

- General business results
- Business process–related results
- Policy-related results
- Organization-related results
- Product- or service-related results
- IT-related results
- Customer-related results
- Supplier-related results

You want to have these items reviewed as well.

Identify Business Issues

Vision, mission, and objectives are all positive. Business issues, on the other hand, are negative and tend to arouse defensiveness. When you are working in this area, be sensitive to the discomfort it might generate and emphasize that you are focusing on the issues so that the process improvement efforts can work to address some of them.

As with the other areas, you will organize these issues into categories:

- General business
- Organization
- Product or service
- Process or policy specific
- IT
- Employee
- Investor, stakeholder
- Customer
- Supplier

You might want to create an issue table as follows. Importance can be rated on a scale of 1 to 5 (1 = low; 5 = very high). The impact is the effect if the issue is not resolved.

Type	Issue	Importance	Impact	Comment
Employees	Low morale	High	Low productivity	More than a process issue
IT	Three critical systems do not integrate	Medium	Excess manual effort	Change requires procedure and systems modifications

Collect Data on Processes

You can also create a list of key processes. You can do this in several ways. The easiest is to take the company's organization chart and write down key processes for each department. This can be reduced to a table made up of two columns. The first is the department and the second is for the process.

Create Business-Related Improvement Tables

You can build a number of improvement tables that will support you in the later analysis of processes.

- *Business vision versus business mission.* Each entry describes how the mission supports a specific element of the vision. This table can reveal inconsistencies and gaps.
- *Business vision versus business objectives.* This table shows how the specific business objectives support the vision. This can show you what business objectives are the most important.

Taking these two improvement tables together, you can see which business objectives are the most important for both the vision and mission. This allows you to restrict your attention to the key business objectives.

- *Business objectives versus business issues.* This table rates each business issue in terms of its relevance to the specific business objective. The purpose of this table is to show which issues are the most critical in terms of their impact on business objectives.
- *Business processes versus business issues.* This table rates each issue in terms of its impact on business processes. You might have to leave many entries blank due to lack of information.

The first table reveals importance to business goals from the top, and the second reveals importance to business processes from the bottom. This combined top-down, bottom-up approach shows the alignment of the processes to the business objectives. It also allows you to create another useful table:

- *Business objectives versus business processes.* This table reveals which of the processes are most important to supporting which goals.

It is important that these tables are reviewed by management and employees involved in process improvement. The political benefit of doing this is to show that you are proceeding in an analytical way and not a subjective manner. Also, getting input early will raise both interest and awareness.

Construct the Organization ScoreCard

You are also ready to create a scorecard of what you have so far (Figure 2.10). The political purpose of the scorecard is to assess each step in process improvement and show areas of potential improvement.

Area	Score (1–5)	Comments
Completeness of vision		
Completeness of mission		
Clarity of vision		
Clarity of mission		
Completeness of objectives		
Clarity of objectives		
Identification of issues		
Consistency over time		
Severity of issues and need		
Learning from past experience		
Awareness of need for change		
Indication of processes needing improvement		
Attitude toward change		

Filling in this table is subjective, no doubt about it. However, the scorecard is useful because it helps you evaluate your own efforts and measure what others are doing. In this case, it also can reveal the readiness of the organization to embrace change.

EXAMPLES

ASC Manufacturing

ASC management was aware that the company had problems getting more completed products out of the production and assembly operation. So this was a focused effort, not a general one. The ASC improvement team initially organized the work into two tiers. This later became unworkable, as too many issues surfaced to the executive committee. So an informal steering committee was organized. The team plunged into the analysis of processes too early so that later people questioned the choice of processes and how they related to business goals. The team then had to go back and create the improvement tables and analysis that were described in this chapter.

Kosal Bank

Management at Kosal knew that it had many issues to address. Yet the company was profitable. Because the managers had no external pressure for change, they could take the general process improvement approach. The three-tiered approach was established and worked successfully throughout the entire project, which spanned more than eight years.

HETSUN RETAILING

Hetsun first tried to treat process improvement through IT. The managers thought that if they could implement modern e-business and other software, things on the process side would sort themselves out. It did not work out that way. The new systems made things worse. External pressure to fix things grew. A general process improvement effort ensued. The major focus was on Quick Wins due to time pressure. After two waves of Quick Wins, management moved on to other areas. That was too bad. Management missed future opportunities and lost momentum for change. Employee morale suffered as well. Altogether, the results were a mixed bag.

LANSING COUNTY

The situation at Lansing was similar to that at Hetsun in that automation was thought to be a panacea. Finally, an organized effort was made to do process improvement in the operations area of the county. This is where 80% of the employees worked. The three-tiered approach was adopted along with the other parts of the method. This worked because the organization was and is very political.

LESSONS LEARNED

- Carefully consider the approach you are going to use for process improvement. Simulate this approach within the organization, and look at it from different viewpoints—from the perspectives of the employee, management, IT, and so on.
- Build initial lists of the various areas that were discussed. When you talk to people informally, show them the lists. Refine your lists and then have them reviewed again along with the tables. This is a proven way to gain incremental acceptance and participation.

PROBLEMS YOU MIGHT ENCOUNTER

- At the start, some people will indicate that this process will fail like other attempts before it. How do you respond? You should indicate that the focus is on Quick Wins as well as longer-term improvement. Also, you should point out that many people will be involved.

- What is the people you want to have on the team are not available due to other commitments? Don't fight this at the start. Try to keep them informed and get their interest. Later you can seek their involvement.

WHAT TO DO NEXT

1. You can start to build the lists that have been identified in the chapter. This will help you see the effort that is involved.
2. Review the approach that was discussed. Identify who might be involved in the committees as well as the action teams.
3. Make a list of issues that you think you might face in terms of resistance to change. Consider the impact of these issues and start to think about how to address them.

CHAPTER 3

Assess Technology and Industry Factors

INTRODUCTION

Observations on Technology

Many years of process improvement under various names and following numerous approaches have brought forth some basic points related to technology.

- Automating transactions makes them more formal and less flexible to change. However, they are also less prone to deterioration.
- When someone attempts to change a process without considering automation and computer systems, there is a greater tendency for the process to revert to its original state or to deteriorate faster. This occurred in the 1920s with industrial engineering. It happened in the 1990s with reengineering and total quality management (TQM). Major or minor changes were made, but many did not last. The changes were not supported by automation.
- Based on the preceding points, the most effective approach to process improvement includes the automation of transactions. The most recent evidence of this effectiveness lies in e-business. For example, once you establish business-to-business procurement with a supplier and it is working, neither side wants to change it.
- A bane of the effective use of technology in processes is the horde of exceptions, workarounds, and shadow systems. In the 1960s and beyond, the purpose of computerization was to automate an entire process. These efforts largely

failed because of the number of exceptions, among other factors. When you develop software or customize packages to fit and handle exceptions, you have to make changes for each exception. This is prone to failure for several reasons. First, there are too many exceptions to address. Second, each one must be addressed separately. There is insufficient time and a lack of resources to devote to doing this.

Process deterioration is a major concern in this book and is one of the reasons why the term "lasting" is in the title. It does little good and potentially much harm to change a process or transactions and then have them deteriorate or revert to where they were.

With this discussion, how should you consider systems and technology in the context of process improvement? Here are some basic points.

- *Guideline 1.* Only consider technologies that can automate and lead to improvements in processes. Other technologies may be interesting but have limited value.
- *Guideline 2.* Focus technology effort and application on the mainstream transactions. Don't be sucked up into the world of exceptions and the rest, except when it is urgent and essential.
- *Guideline 3.* Do not aim to replace many major systems with the excuse of process improvement unless doing so is the only way to carry out improvement. There are a number of reasons for this. First, system replacement requires major effort that can delay and derail process improvement. Second, who is to say that the new systems may not cause more problems than you already have?

It is clear that you should involve systems in your process improvement effort for the benefits given. However, this is a two-sided coin. Involving systems and technology means that another organization in the company is going to be drawn into process improvement—information technology (IT). Any group, including IT, has its own agenda of work. You will have to negotiate with IT for support. Next, involving IT will raise political problems since other groups must now interface with the IT group. We point these things out not because you want to avoid involving IT—you cannot. Instead, you must realize that the support of the IT group is often critical for success.

When you carry out change and seek to make changes in the technology and systems base, you are bound to run into some rather long-standing problems and issues. Let's suppose that you want to implement internal intranets for e-business internally. Upon analysis you find that the network is not up to supporting this workload. What initially seemed to be straightforward now involves a major network upgrade. It is the same when you repair something around your house or apartment. You get started and find other problems that cannot be ignored. That is why you need to identify technology-related factors early—to see what you are in for.

Remarks about Industry Factors

Let's now turn to industry assessments and studying the competition. We are trying to carry out process improvement. This is, in many cases, mostly internal. Why do we need to consider industry and competition? We already have enough to do! The answer to this question is twofold. First, there is the negative side. If you don't understand what other firms in the industry are doing, then you risk doing all of the work only to find that you are still behind. It was this way with a major retailer who attempted to carry out improvements but never came close to the position of Wal-Mart. Yet what Wal-Mart was doing was out in the open. The other retailer had the opportunity to use the information to get more innovation and blew it. Another aspect of the negative side is that people will find out what others are doing and raise issues. You will be put into a defensive, reactive position. Everyone from Genghis Khan to Machiavelli has said that this is not good.

Now, for the positive side. If you make the effort to gather information on what firms are doing, you can get some positive ideas on what processes to go after. You can learn from other people's mistakes. With Six Sigma methods, you sometimes tend to focus too much on internal issues.

E-Business

The preceding comments all come together in e-business implementation. What are you doing when you implement e-business?

- You are creating new and changing existing internal processes. These internal processes and transactions link to suppliers and customers. Traditional work has an internal focus.
- You are outwardly facing toward customers, suppliers, and employees. Traditional change such as reengineering has focused on internal change.
- You are automating transactions internally and with suppliers and customers. Automation in the past focused on what the organization wanted.

These factors point to a critical difference from systems and improvement efforts of the past. This difference is where the center of the change universe is. For e-business, the center is the process and the customer, supplier, or employee. Traditionally, for the past 40 years, automation has been centered on the organization.

OBJECTIVES

Technical Objectives

Technology and industry assessment and analysis share the common objective of uncovering potential problems and opportunities. In the area of systems and technology you are trying to do the following:

- Determine the problems and issues with your current systems and technology that will impede process improvement.
- Identify potential new systems and technology advances that will facilitate and support change and the new transactions and processes.

Moving to the industry assessment, you are attempting to do the following:

- Find out what other companies have as their goals with respect to the business so as to help you select the most appropriate processes for change.
- Ascertain what methods and tools various firms have employed that they feel have been successful.

Business Objectives

There are several shared business objectives for the technology and industry analyses:

- You want to involve employees and managers in understanding the potential benefits of what others have done and what new technology and systems can do.
- You can use the results of the analysis to get support for process improvement. This is very beneficial in cases where the company is complacent because it is doing okay from a profit or business view.

Political Objectives

One political objective is to raise the need for urgency. The company might fall further behind unless it modernizes its processes and systems. Another, more negative objective is to increase the fear factor—being left behind and all of the consequences this has for people.

A third political objective is that you want to build grassroots support for process improvement. Sure, you can rely on management to get process improvement started. But management eventually turns to other pressing problems. After all, the processes are working. Process improvement can be viewed as an enhancement, not a necessity. Thus, in the end it will tend to receive less management attention even though it is important. You must build grassroots support for changes in processes and for preventing process deterioration of the new work methods. This is a critical difference between Six Sigma and reengineering. In Six Sigma, you involve many people. In reengineering, the work is typically performed by a team with less involvement by lower-level employees.

END PRODUCTS

There are more end products than you think. This statement will be true for each chapter. The obvious end products are as follows:

- Identification of problems with the current technology
- Understanding of the impact of the systems and technology problems on the processes and work
- Knowledge about trends, successes, and failures in the industry
- Application of this business knowledge to the process improvement effort

Now let's consider other end products. You want people to understand that if the critical problems with the systems are not addressed, the degree of improvement and, hence, benefits will be limited. The end product is that people see why this additional change is compelling.

In terms of business, the end product used is to jar people out of their complacency and isolationist views. Instead of being internally focused, they become more aware of what is going on outside and what is possible.

The next end product is to instill an attitude among managers and employees that they have to be aware of what is going on around them on an ongoing basis. Of all of the end products, this is the most lasting and, perhaps, the hardest to achieve. How you go about the analysis and assessments determines the degree of your success.

METHODS

Where to Start

Remember that we have assumed that you have limited time and resources. We are assuming that very little time and energy can go into this part of the work. We combined these two areas into one chapter for a purpose—to achieve economies of scale.

You should start by using the Web and magazines to look at how other companies are doing their work and using technology. Appendix 3 lists a number of magazines. A substantial number of these are available without cost through the Web. Set up a new e-mail address at one of the service providers such as Yahoo or Excite. Enroll for as many of these magazines as you can. This will start getting you information on a regular basis. Next, when you read the newspaper or popular magazines, look for stories of success and failure.

Here are some specific things to look for:

- Consultants and firms often try to get stories about their work with clients published. This helps them get more new business. Look for these stories and be aware that there is the hidden agenda—the stories are not likely to be unbiased.
- Stories about how companies applied systems, software packages, and technology abound in the technical-management literature. These are useful because they include (1) benefits, (2) the combination of the technology that was employed, (3) how they measured success, (4) how they went about the work and organized it, and (5) lessons learned from the efforts.
- Stories and reports by analysts about companies that have achieved success or experienced failure. From these you can extract both success and failure factors.

We are not done yet. Go around the company and test the level of awareness of the outside world. Find out what new technologies the IT organization is exploring and considering. Find out what concerns the business area has. Do this informally. Here is a list of questions that you want to answer:

- What technologies have been and are being considered?
- What firms do business managers hold out as examples of success?
- What are some of the problems that people see as significant? This is piggybacked on the work you did in the previous chapter on understanding your business.
- Suggest some examples of business success and failure as well as technologies and see how people respond. What is their reaction? Not only can this reveal additional data on the value of this information, it can also tell you a lot about people's attitude toward external information and, indirectly, change.

Remember that you are doing the work in this chapter in parallel with the work in the previous and following chapters. You have to do this because of the time pressure.

Examine Your Current IT and Systems

Starting in the 1970s, firms began to understand that what counted was how the technology and systems went together. More than the individual components, it was the whole that counted. The structure of the systems and technology was called the architecture. Use of the term and analysis of architecture have continued ever since.

Try to get hold of reports and presentations on technical architecture. This can often reveal potential problems and limitations. It can also indicate what things are a priority in terms of architecture.

An IT group usually publishes presentations and reports on its activities. These consist of annual plans, active projects, and progress reports. Review these documents to find out what is currently going on. You can use them to identify IT objectives, projects, and issues that relate to IT.

So what do you do with the information? You must organize it if you want to use it effectively in your process improvement effort. Start creating several lists. These include the following:

- Key technologies in use for the network, hardware, and system software
- Major application systems and software packages
- Architecture components at a high level
- Issues and problems that IT indicates must be addressed

These lists will serve as the basis for improvement tables later. You can have people review the lists. If they ask why you are doing this, answer that you are attempting to gain a better understanding of the systems and technology in use so that you can understand the feasibility of process improvement from a technical, practical view. To give an example of potential issues, consider the list from our past efforts that is shown in Figure 3.1.

Assess Potential Systems and Technology

There are many potential technologies available at any one time. It is tempting to be drawn into this stuff. It is interesting! Be careful and be selective. Figure 3.2 gives a potential list of technologies and some of their applications. Make your own list. You will later use this list to see what impacts the new systems and technologies can have on the processes and work.

How do you find these technology and systems candidates? Go to the literature. Visit vendor sites. Here is an example of one technology: Microsoft and several firms have developed software that acts to support interfaces among systems.

- Network performance is quite slow, inhibiting potential new applications.
- There has been staffing attrition in IT, leaving holes that are difficult to fill.
- Some of the application systems are so old that maintenance and enhancements to support process improvement are difficult to carry out.
- The current network is quite dated and is composed of a variety of old technologies.
- The IT group attempted to implement new systems recently and these efforts failed.
- It takes too long for even routine changes to be made to the software.
- Many departments have developed and currently use their shadow systems.

Figure 3.1 Some potential issues with current systems and technology.

Microsoft calls its software BizTalk Server. The software typically has the following features:

- You define your own transactions through wizards.
- You then define the format and structure for other systems, including those of your customers and suppliers.
- After setup, the software will take transactions from one system and then restructure them for another system.

Hardware and system software
- Raw hardware processing (e.g., 64 bit computing)
- Operating systems
- Mobile devices
- Data base management systems
- Object oriented software
- Security software

Networking and communications
- Network operating systems
- Network management software
- Network hardware
- Network performance tools
- Electronic mail
- Collaborative software

Application software
- New versions of products that you use
- New services from ASP's and others

Figure 3.2 Potential technologies of interest in business processes.

To a nontechnical person this sounds very simple and straightforward. However, it has been shown to yield major benefits in e-business transactions and in regular IT work. Here are some of the benefits cited.

- You can develop interfaces much faster than with the traditional approach of custom coding each interface individually. These are called application program interfaces (APIs).
- You are likely to do less testing, since the software is stable.
- You can consider a wider range of interfaces than you normally would due to schedule and resource limitations.

Many firms have employed this type of software. They have used it to interface newer software and older, legacy systems, for example.

How can this software assist process improvement? To improve processes you sometimes want to interface and integrate various systems. If you went the traditional route, it would take too much time. Who is to say that you would even get the resources to do it?

Now let's step from this detailed example to a better understanding of how systems and technology are used in a company. This will help you better their role in processes.

Technology and systems can be divided into several categories: core, niche, and temporary. Core technologies are those that the organization is dependent on. Examples or core technologies are the network, hardware, and major application systems. Niche technology is a technology that fills a specific need for a specific process. An example of a niche technology is wireless radio frequency, bar code readers for a warehouse. Both categories contain systems and technologies that are longer lasting. As it's name implies, the third category is more temporary. Temporary technologies deliver some benefits, but will have to replaced over time. Most new technologies fall into this category. Later, when you receive improved versions, they might go into the core technologies category, or if they are used for a special purpose they become niche technologies.

Now let's see how this plays out over time. At any given time, a specific technology is either not used (being potentially useful or irrelevant), core, niche, or temporary. Over time, new technologies are used and older technologies disappear from use. For example, Figure 3.3 shows two diagrams. In the first you see five technologies: 1 is core; 2 is niche; 3 is temporary; 4 is potentially useful; 5 is not relevant. In the second, you can see that 4 became core and 3 disappeared. This shows how technology changes over time.

What does this have to do with process improvement? You will be evaluating current and new systems and technology. With the IT group you will be making decisions and implementing changes. How the technology fits is important. Here are some observations:

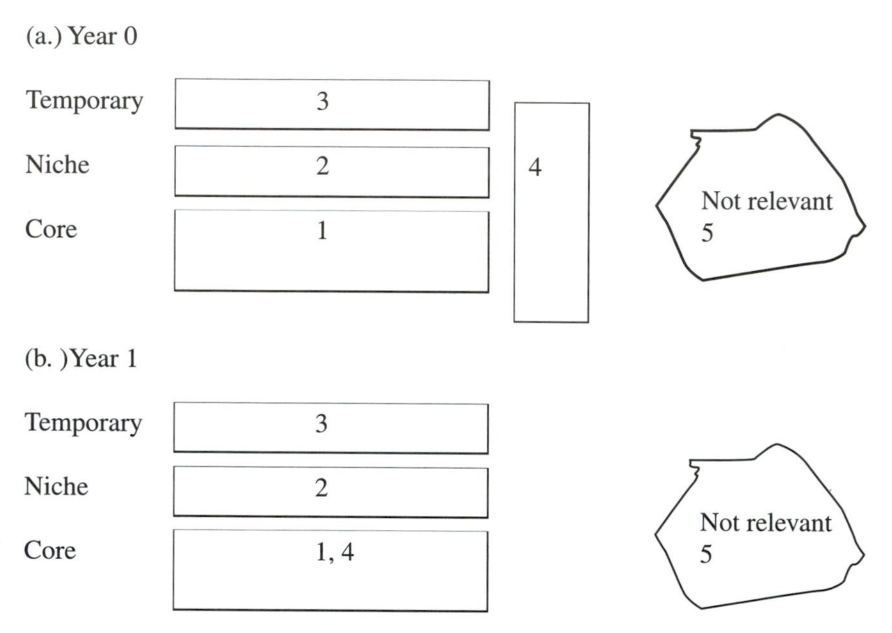

Figure 3.3 Examples of technology categories.

- You do not want the new process to depend on some new temporary or transient technology. It may later evaporate and leave you and the process high and dry.
- Core technologies tend to be used by many people. If you make changes in a core system for process improvement, it can possibly yield benefits for many people—a good marketing feature when you sell process improvement.

Construct the Improvement Tables and the IT Scorecard

Remember that this is a limited effort; you are not attempting to fix IT. However, if you were, you could use an expansion of this approach to deal with IT. Here you will construct a summary scorecard and some improvement tables. Recall that in Chapter 2 you used critical processes, mission, vision, objectives, and business issues. Now you can add architecture components, IT objectives, IT-related issues, and potential systems and technologies to create new improvement tables. Space does not allow us to list all combinations of tables. Here are some of the better ones to use. We begin with IT-related factors. Note again that you are not attempting to be complete. Rather, you are building up the case for process improvement and getting an understanding of what you will need. In most of the tables you can employ a scoring system of, say, 1 to 5, in which 1 is very low and 5 is high.

- *IT objectives versus architecture components.* Here you enter the degree to which the IT objectives address architecture. This is useful for two things. First, it shows which components IT wants to address. If they are the same as what you later find you need for process improvement, you have demonstrated alignment with IT. Second, you can see how focused the IT department is on its architecture.
- *IT objectives versus IT issues.* Here you enter the degree to which the individual objective addresses the specific IT issue. This reveals which IT related issues are viewed as being important. Your hope is that process improvement will address some of these same issues.
- *IT architecture versus potential technologies and systems.* The entry is the goodness of fit of the specific new technology and the architecture component. This reveals how likely it is that the IT department will adopt a new technology of interest to process improvement.
- *IT architecture versus IT issues.* The entry is the impact of the issue on the architecture and whether the architecture component has that issue. This table helps to show which issues are important from an operational point of view.

Now we can use these IT lists and the previous lists in Chapter 2 to create the following additional improvement tables.

- *Critical processes versus architecture components.* The table entry is the degree to which a process depends upon the specific architecture component. This is valuable since it shows what parts of the architecture are important across many processes.
- *Business issues versus IT issues.* This table addresses the question, "To what extent do IT issues align with business issues?" Your hope is that the process improvement effort will address significant issues in both categories.
- *Business issues versus architecture components.* This table indicates the extent to which a business issue depends on a specific architecture component.
- *Critical processes versus IT issues.* The entry in the table is the degree of impact an IT issue has on a critical process. The purpose of this table is to reveal the impact of specific technical issues on the processes.
- *Critical processes versus potential systems and technologies.* This table shows in a general rating the potential impact of something new on a process. Note that like all other improvement tables, the entries are subjective and rely on what you gathered so far. As such, they are subject to later change.

After you prepare the tables for the first time, they can be updated and eviewed as you proceed. Since process improvement is a program and not a one-time project, the improvement tables will also be of long-term use and value.

How can you use these and others that you will prepare? Just as in Chapter 2, you will be presenting these to management to get feedback to ensure you are on the right track. You will also be supporting your political objective of getting people involved.

Now you can move to the overall IT scorecard. Scorecards for IT like many other areas are dependent on the purpose. Here your purpose is to assess the readiness of IT to support process improvement (Figure 3.4).

Evaluate Your Industry and Competition

Now let's move to the industry and competition. It is difficult here to develop tables since the type of information that will be available cannot be predicted. First, you will want to organize sources with copies of the articles and web pages. Here are some categories for organization:

- Approach to process improvement employed
- Issues faced by the competition
- Specific processes employed by other companies
- Technology and systems applications
- New technology and systems, examples of applications
- Suppliers and their technologies
- How firms measure their processes

Element of the Scorecard	**Score**	**Comments**
Dependence of processes on IT		
Relevance and impact of IT issues on processes		
Degree of potential architecture and technical problems		
Backlog and workload of IT		
Alignment of IT to business processes		
Degree to which business issues are technical problems		

Evaluate Your Industry and Competition (H3)

Figure 3.4 IT readiness scorecard.

You can find out a great deal about a company by going to a search engine and searching for the company. When you get to the financial data, look for comments by investors and others. You will find some very frank opinions. Some can be very nasty. These, you should remember, are only people's individual views and are not substantiated.

Who has time to do all of this work? Here is a creative approach that we have used in several firms. Obviously, managers, IT staff, employees, and you do not have time for doing this on an extended basis. Who has the available time? There are two groups to focus on. One consists of the parents of employees who are retired and the other is the children of employees. Both of these have time. How can you motivate them? Run contests and give away certificates, gifts, and other items. Perks like these can raise morale.

Create the Argument for Process Improvement

In the preceding chapter you analyzed the internal mission, vision, objectives, and processes. In this chapter you have now assessed both the technology and industry. You have sufficient information to go to management with more ideas on process improvement. Up until now we have assumed that either there has been no formal approval or that the approval has been rather tentative. We did discuss how to organize the effort. However, even if management is enthusiastic, you have to make the effort to reinforce the support.

Let's consider how you can make the argument for process improvement. First, it is best if you have a target in mind. E-business is a good target because it is positive and different. Focusing on cost-saving processes like reengineering can really drag down morale and have a low chance of success.

Without a target, you would to have compelling need. Some examples of needs were pointed out in the preceding chapter. With the additional information in this chapter, you can now get more precise. Here are some specific actions:

- Use the improvement tables and information to discuss the business issues as well as IT issues. Show how process change can help resolve some of these issues.
- Give examples of firms who were in trouble and did not improve things and then indicate what happened to them.
- Discuss other examples of success where changes were made in industries similar to yours.

What is your goal? You are trying to get people to move to the next step. That is to select the processes and transactions for improvement.

EXAMPLES

ASC Manufacturing

In order to implement the new scheduling and work control processes, the company needed new software tools and methods. An action team of business and IT staff conducted an extensive search. There was no complete solution from one supplier, so the effort then turned to what could be accomplished by a combination of off-the-shelf technology. This alternative later proved successful. After two years, the off-the-shelf technology was then replaced with more modern tools. You could say that ASC adopted the technology on a temporary basis in order to get the new processes implemented.

Kosal Bank

Kosal had many problems with its existing architecture. It became clear after analysis that a new network was necessary. The industry assessment then identified several approaches that could be taken. Kosal picked one of these approaches based on the industry assessment. This example shows how the technology and industry assessment go hand in hand.

Hetsun Retailing

Hetsun Retailing had many internal systems issues. There was inadequate staffing. The previous approach was to keep IT costs very low. Hence, there were always performance problems and system crashes. Central to process improvement was the effort to fix these problems. Not only were additional resources and staff needed, but also much effort was devoted to changing the attitude of management toward technology. This required a number of meetings and discussions at which the tables that have been described were presented. Eventually, management adopted a more positive tone toward IT, but the long battle consumed precious time.

Lansing County

Lansing had just gone through a major upgrade and migration of its core systems to new hardware and network structures. The effort was successful. However, both the IT staff and management were tired and did not want to pursue new initiatives. To move process improvement ahead, it was necessary to base much of the technology and systems effort in a major business unit.

LESSONS LEARNED

- Measure your progress not by the volume of data that you have collected, but by the holes in the information that remain. This is a more demanding and useful standard.
- Be wary of just-announced technology. It takes time for new technology to settle down and for technology manufacturers to iron out the errors. It also takes time for people to gain experience with a new technology and determine how best to use it.
- To assess the benefit of a new technology, assume that you have to modify the process and transactions without the new technology. What additional efforts are needed? This will help identify the benefits of the technology to the business process.
- Avoid getting a deep understanding of the new technology. There is no time to do this and it diverts you from other activities.
- Look for the hidden costs of maintaining and operating the systems and technology in production. We have seen cases where a new technology was just perfect for a process, but the application failed because too much support was required.

PROBLEMS YOU MIGHT ENCOUNTER

If your experience is like ours, you will run into skepticism. Here are some of the doubts people might express and some tips on how to respond:

- "We don't have time to do process improvement with our regular work."—Our experience is that if you organize the effort following the approach in the preceding chapter, you will be able to do the work.
- "Taking on processes is too much work and too vast a scope."—This is true. Your response is that the effort will take on selected transactions, not the myriad exceptions and workarounds. You will only address these issues when it is critical.
- "IT work is such that resources are not available."—Here you can to the tables and lists. You will be indicating that you will seek out tools such as the one in the example. You will also meet with the IT staff to make plans and set schedules.
- "We have tried improvement efforts in the past, and they did not work."—This is often the case. Remind those concerned, however, that this approach is different because it uses common sense and assumes a limited effort.
- "Even if we make changes, they will not last. The work will just go back to the way it was."—You can respond to this concern by the use of

automation for selected transactions. Also, there will be ongoing measurement and tracking to detect signs of reversion and problems.

WHAT TO DO NEXT

1. It is useful to construct a diagram of the current architecture. When you have done this, show how the IT and business issues relate to the diagram. This brings home to managers who are not familiar with IT some of the problems and impacts.
2. Once you have identified new technologies, you can now construct a new diagram of the potential future architecture. Then you can compare the diagrams from the current technology and the potential new one and show how the new one is better.

PART II

Define the New Business Processes

CHAPTER 4

Select the Right Processes

INTRODUCTION

In Part I, you gained a better understanding of the business environment—both internally and in the marketplace. It is now time to put this knowledge to use. This chapter is important because it sets the stage for everything that follows. If you pick the wrong processes to improve, then your entire effort may go down the drain. Within the processes selected, you will want to select the transactions and work that account for the bulk of the activity. Remember the guideline to try and avoid being mired in exceptions and workarounds.

Here are some examples of failure:

- A garment industry firm selected more than 60 processes to improve. There was little cooperation among the teams doing the work. The firm almost went bankrupt.
- A pharmaceutical distribution company decided to address several processes at one time. However, these did not interrelate with each other. The impact was that any benefits were outweighed by problems generated in surrounding processes.

Here are some mistakes that you can make:

- *Center your attention on one process.* Processes are interrelated. If you touch one process, you typically affect the surrounding processes. Moving customer ordering to the Web for e-business will then affect customer order tracking, for example.

- *Spread the effort to too many processes.* Not only is the effort diluted, but also the increased coordination effort can spell disaster.
- *Address entire processes.* This sounds like a good idea at first and is common in Six Sigma and other methods. Unfortunately, when you follow this approach, you are likely to be sucked up in coping with exceptions, workarounds, and shadow systems. While you have to deal with some of this, you can get carried away. As time goes on, more exceptions appear. The work will never be done if you strive for completeness.
- *Copy what other companies do.* This sounds good at first. You find a competitor or similar firm and you find out what processes they addressed. So you do the same thing. It might work and it might not. Without analysis, you probably will, at the least, miss some opportunities.

These simple examples reveal the importance of selecting the processes right. In many applications of Six Sigma, the focus is on the customer and customer requirements. A key process is selected based on its importance to the customer. While the Six Sigma goals are common to many methods, the customer-only focus is too narrow and restricting.

You also have to be concerned about the scope of what you can do. Almost all improvement methods focus on improving a process in place. This is, however, not true in e-business or outsourcing. There are several options. Here are four:

- Improve the work in place (overlap). Here you improve part of a process.
- Move work to new processes and away from the current work (separation). An example is when a firm establishes an e-business subsidiary.
- Combine and restructure work (integration). This is more comprehensive change.
- Totally replace the current process with the new one (replacement). This option would fit reengineering.

Figure 4.1 provides a diagram for each of these strategies. The square is the current process, and the circle is the new process. As you can see, there are only four possible ways to arrange a circle and a square relative to each other. This reveals that these four alternative strategies are comprehensive.

Now let's turn to defining process selection. You are not just going to select a group of processes. You are going to select transactions within these processes. Remember that you are under time pressure to show results. This is not an academic exercise. The more transactions you select, the more analysis you have to do and the longer it will take to implement change. This means more risk. However, on the other hand, if you select too narrow a range or scope of work, then you risk not getting any meaningful results. It is a trade-off.

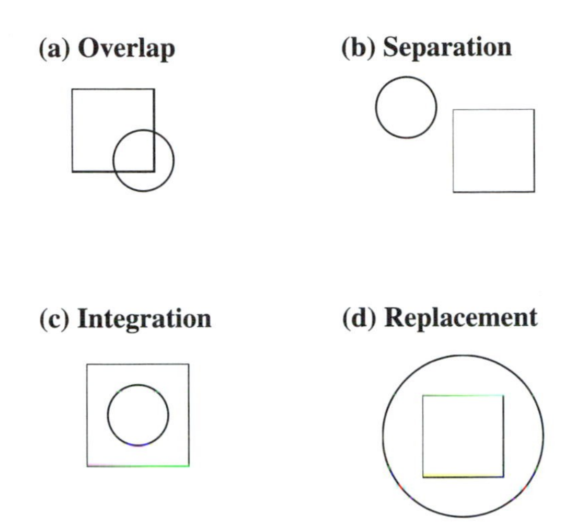

Figure 4.1 General strategies for process change.

OBJECTIVES

TECHNICAL OBJECTIVES

The technical goals are fairly obvious. You want to select a combination of transactions and processes that can be improved with reasonable effort in limited time. You would like to select the work so that you can improve your skills and gain support for further improvement efforts.

BUSINESS OBJECTIVES

The business objective is to select processes and transactions that yield the most benefit to the organization for the amount of effort. Another business goal is to limit risk to the business. Thus, cross-impacts among processes are important.

POLITICAL OBJECTIVES

Many political objectives cannot be ignored. These include the following:

- Gain support for the processes and transactions that were selected. Otherwise, people keep asking, "what if?"

- Gain management and employee support for the changes in the processes.
- Pave the way for improvement in other areas.
- Select work where you can show some Quick Wins and results. Selecting processes that can only be improved with a massive automation effort, for example, will often lead to failure.
- Demonstrate that you and the team have the capability to carry out change and improvement.

END PRODUCTS

Considering the various objectives, you can work toward the following end products:

- Selected processes and to an extent the desired transactions
- Awareness of why the processes not selected were dropped
- Support for the improvement effort

METHODS

Where to Start

You will be selecting a group of processes, rather than one process. Why is this emphasized? Because in many situations you are dealing with a customer, supplier, or employee and looking at the process from that party's perspective. That means you are looking at processes from a view different from that of the internal organization.

Determine Process Candidates

Using the work from Chapters 2 and 3, you will first make up a list of potential processes. Let's consider a new example—an insurance company that wants to get into e-business with customers. Figure 4.2 gives a short list of processes from the customer point of view. Notice that the internal accounting processes are not listed.

With a list of processes, you are prepared to do some initial analysis of the processes. The first action is to determine how the processes relate to each other. If they turn out to be very dependent upon each other, then you will face a larger and probably more complex implementation effort. Obviously, no two processes that relate in any way will be independent of each other.

- General insurance product information
- Insurance application form
- Insurance application processing
- Status of application
- Issuance of policies
- Payments and insurance premiums
- Review of insurance status
- Name, address, and other maintenance
- Billing
- Claims
- Cancellation
- Change of coverage

Figure 4.2 Potential list of processes for the insurance example.

How can two processes interface?

- *One process passes instructions to the other.* A process can depend on another for procedures as well as information. This is the strongest type of process linkage. This occurs, for example, in handling exceptions.
- *The two processes share information.* Here the first process feeds data to the second process as it goes along. It does not completely finish the work. Some processes take substantial time to complete their work. If you waited until this process was complete, the overall transaction time would be excessive. This is the case in a hospital where a staff member reviews the patient information at the same time that diagnosis is started.
- *One process depends on the completion of another.* In order to do insurance billing, you have to collect the information on the insurance application and have performed an insurance rating to determine the cost.

These are just three examples. You can find cases of many variations. Two processes do not have to interface with each other directly. They might relate to each indirectly through infrastructure, organization, and resources. Here are some examples:

- *Processes share hardware and the network.* Both processes operate on the same IT infrastructure. This can be good, since it shows that they have a level of technical compatibility. It can be bad if the capacity of the hardware and network are limited so any changes you make to one process affect the performance of the other.
- *The processes share facilities.* Then making procedural or structural changes in one process can impact others.
- *The processes share the same equipment.* Examples might be machinery in a manufacturing or assembly operation. Changing one process may require

new and different equipment—affecting all processes using the existing equipment.

- *The same people perform both processes.* You can make changes to one process through procedures and then confuse people. The overall performance can become worse.
- *The processes affect the same suppliers, employees, or customers.* Here making a change to one process may lead to lost business or discontent among suppliers or employees.

When you consider processes, you will want to group two processes that are closely linked or connected. In the insurance example, you might link billing and payments, for example.

Identify Transactions

You can use the analysis of how processes relate as well as the goals of the effort to determine which transactions to consider in each process. In the insurance example, the company offers automobile, mobile home, and homeowner's insurance. These are their three key products. Since they are just getting into e-business, it is logical to concentrate on the simplest and most frequently processed product—automobile insurance.

Define Alternative Groups of Processes

You are now ready to generate alternative groups of processes. Here are some criteria for generating a group.

- *Customer based.* Group the processes that touch a particular aspect of your customer relationship. In the insurance example, you could group application forms and application processes. You could add billing as well.
- *Supplier based.* Let's suppose that you purchase materials and supplies from many different vendors. You might group processes by those that relate to changes to purchase orders, for example.
- *Function based.* Group all processes that perform the same function. This is one of the safest groupings, but it will probably increase complexity by spanning multiple organizations. For example, you might group application processing across all insurance products. Each product is supported by a different group.
- *Organization based.* Group all processes that the organization operates and manages. This was the focus on many industrial engineering efforts

where participants worked in one group to get improvements. Note, however, that the placement of a process in a specific group may be the result of politics rather than logic.

- *Technology based.* Here you group all processes that use the same technology. In banking you might consider all processes that employ an automated teller machine (ATM). This grouping is desirable when you are improving by enhancing the network.
- *General manager based.* All of the processes under a specific high-level manager are grouped together. This was a classic reengineering approach in which you worked in the division or organization of a high-level manager who championed reengineering. This is mostly a political grouping and may make little sense otherwise.
- *Financial performance based.* When a company is in trouble, you might want to center your attention on the processes that are losing money and are highly inefficient. This happens when you are restructuring a company or when you are getting ready to sell off assets.
- *Business objective based.* You begin with one of the business objectives in Chapter 2 and then identify processes that relate to this objective. It draws directly from the corresponding table.
- *Competition based.* You examined what competitors and other firms in the industry are doing in Chapter 3. Here you will select the processes that they changed to achieve success and competitive advantage. That is what some retail companies did in copying Wal-Mart or banks that copied Citibank. This is not an easily sustained grouping since it is not natural to your organization.
- *Issue based.* You identified business issues in Chapter 2. You will select processes that relate to a specific critical issue. An example might be poor customer service as an issue. Then you would select all processes that touch on customer service.
- *Quick Win based.* Here you would select processes based on where you could get some rapid results and improvements. You might end up picking a strange grouping. Later changes may also undermine what you have done here.

It is useful to employ graphs and charts to help you and others understand what is going on. Figure 4.3 is a spider or radar chart. It is useful to us because you can show many dimensions in one chart. Note that the actual values of the graph will be subjective. Your political purpose in using such charts is to generate discussion and get others involved in process selection. In this example, the following are the dimensions:

- Number of organizations involved (dimension 1)
- Number of high level managers involved (dimension 2)

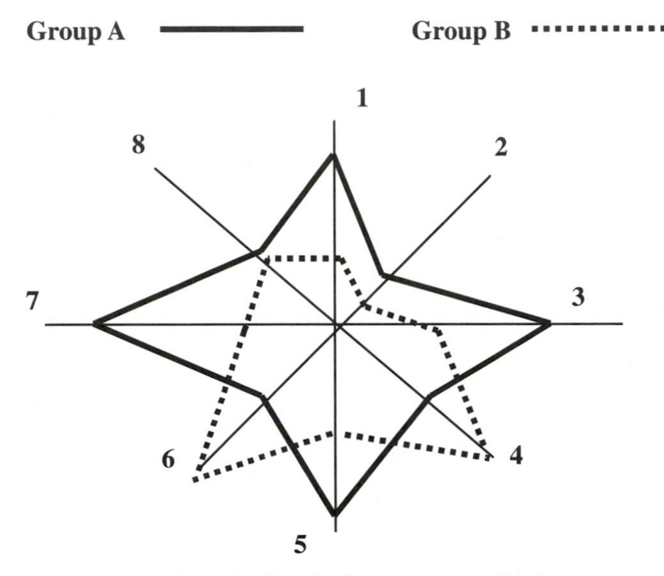

Figure 4.3 Example of graph of two groupings of business processes.

- Number of employees in affected departments (dimension 3)
- Number of facilities (such as locations) involved (dimension 4)
- Total volume of work performed (dimension 5)
- Number of different transactions involved (dimension 6)
- Number or percentage of customers or suppliers impacted (dimension 7)
- Number of software application systems involved (dimension 8)

In the graph, group A is high in organizations, number of employees, and number of customers. Group A is moderate in number of transactions, number of systems, and number of facilities. Group B, on the other hand, involves fewer organizations, fewer employees, and more transactions. From this initial analysis, it would appear that group A is a better candidate since it has a greater impact but may involve less work since there are fewer transactions.

These dimensions are useful for showing the scope of two groups of processes (A and B). You might reject groups that were too limited because there may be insufficient benefits. You also, at the other extreme, might reject groups that are too wide in scope. Change here may be too complex.

The typical situation for a group of processes is to select one or two to be the core of the group. You will then add smaller processes that relate to these to complete the group. Examples of these are as follows:

- Processes that support letter generation, e-mail, or faxing from output of the primary processes

- PC-based shadow systems that use output from the primary processes
- Electronic forms that link to the core processes in the group
- Groupware that can perform collaborative support
- Manual steps and processes that are extensions of the process

Be careful here that you do not take in too much. This is a primary time when exceptions, workarounds, and shadow systems can creep into the group and make it unmanageable. If you have too many processes in the group, you can become bogged down in infrastructure change, analysis, and other activities. The number of reviews, meetings, and communication activities increase. The chances of generating more political problems increase.

For the insurance example, consider Figure 4.4. The first group pertains to customer service and the second to claims. Note that since most customers do not file claims, there is limited impact.

Select the Right Process Group

In order to select the appropriate process group, you can prepare more improvement tables.

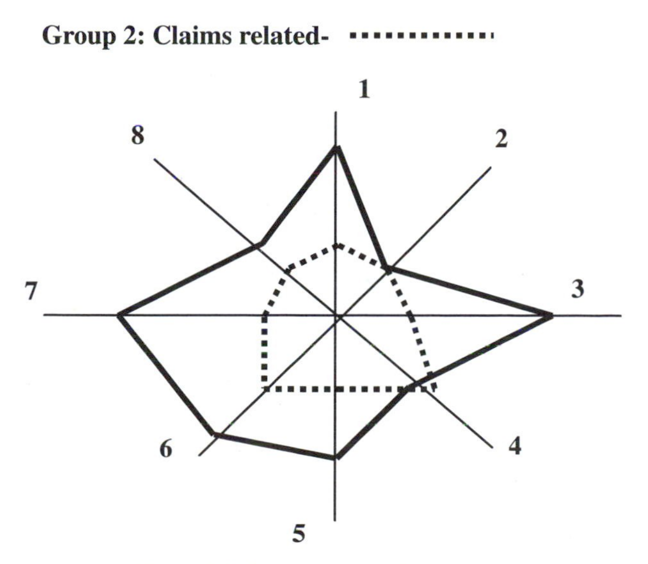

Figure 4.4 Insurance example of two groups.

- *Process group versus process group.* The entry is the relationship between the processes in each group. This table allows you to discuss the individual processes in each group.
- *Process group versus architecture.* The entry indicates the degree of dependence of the individual process on the architecture component. You can use this table to assess how much of the architecture is impacted by the processes in the group.
- *Process group versus business issues.* This table shows how the processes contribute to the various key business issues. The table entry is the degree to which the issue is due to process problems.
- *Process group versus organizations.* The entry indicates the degree of involvement in the individual process. This reveals the scope of the improvement effort likely for each group.

Suppose that you have identified a number of alternative groups. In the insurance example, you could identify groups for applications, customer service, billing and accounting, and claims. You do not want to pick one group yet. You want to narrow the field and get management involved in the selection. Based on our experience, you should winnow the field down to two groups. Here are some actions to take:

- Develop additional tables for each group. You can use more of the factors identified in Chapters 2 and 3.
- Determine conditions under which each group would win in the competition. Here you can consider political factors such as the power positions of various managers.
- Proceed by the method of elimination. You could, for example, eliminate all groups that do not support the key business objectives.

Develop Process Group Scorecards

Once you have narrowed the field somewhat, you can consider the following categories of evaluation:

Performance
- Revenue and costs: Your assessment of the potential impact of changing the process group on revenues and costs
- Competitive position: Your view on how changing the specific group might impact the competitive position

Internal
- Organization: The potential impact of the process changes on the organization

- Infrastructure: The possible effect of process change on infrastructure and the IT architecture
- Impact if not done: An assessment of the potential effect if the changes are not undertaken (the opposite of benefits, so to speak)

External

- Customer: The potential benefits to customers of changing these processes
- Supplier: Your assessment of the impact of process change on suppliers

Implementation

- Ease of implementation: An assessment of potential resistance and problems you might encounter if you selected a particular group
- Risk: Includes risk to the business, processes, and the improvement effort itself
- Elapsed time: An estimate of how long it will take to implement the changes for the specific group
- Estimated effort: The estimated labor effort for implementation of improvements
- Availability of employees: The degree to which you think employees will be available to work on the improvement effort

These factors become elements of scorecards for process groups. Using these scorecards for the processes, you can narrow down the field to two process groups. You can employ an evaluated spider chart for the last two (see Figure 4.5). These are the same groups used in the previous diagrams for the insurance company. Claims improvement tends to be more complex and have fewer benefits due to the extent of individual business rules and the existence of many exceptions that have to be addressed.

How do you go about doing this analysis and estimation? You will want to employ the project template to give you consistency. Figure 4.6 provides an example of a template for implementing e-business. This one has been included because many improvement efforts are centered around e-business implementation. The areas of the plan in this figure follow the method in the book.

What else can you do for evaluation? Let's turn to advertising. An advertisement on television always shows how good you will feel or how happy you will be if you buy and use the specific product. You can use this proven method here. For each of the two process groups, think about how the world would be if the changes were made to each group. Answer the following questions:

- What would customers or suppliers do differently?
- How would employees perform the work differently?
- What additional information would be available for management?

You can now develop an overall scorecard for how you are doing in process selection (Figure 4.7).

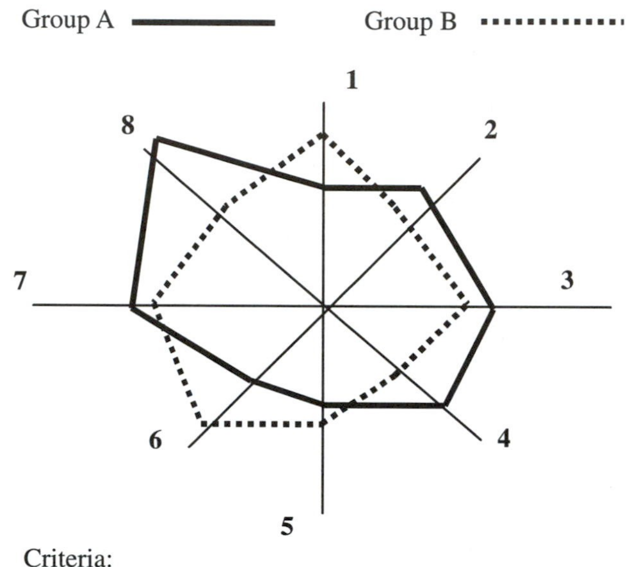

Criteria:

1- Risk
2- Internal benefits
3- Customer/supplier benefits
4- Cost impacts
5- Ease of implementation
6- Elapsed time
7- Effort required
8- Impact if not done

Figure 4.5 Evaluation of two process groups.

Gain Management Approval for the Selection

You are likely to run into several management questions and issues. Here are some common ones that we have encountered:

- Assurance that the choice of processes is correct and the best under the circumstances. This is in part due to management being aware of many failed efforts in the past.
- Concerns about processes that will not be improved. What will be done about these? What is the political fallout?
- Worries about resources. Since everyone is doing other work, where will the people come from to carry out this effort?
- Concerns about the attitude and reaction of employees about change and the impact on their jobs.

- 1000 Overall e-business vision, objectives and plan
  - 1100 Develop e-business vision
  - 1200 Define the e-business objectives
  - 1300 e-business management
    - 1310 Identify issues management approach
    - 1320 Identify management approach
    - 1330 Establish project steering committee
    - 1340 Establish the e-business executive committee
    - 1350 Identify project leaders
  - 1400 Identify organization roles
- 2000 e-business strategy
  - 2100 Define overall e-business goals
  - 2200 Determine mission of organization with e-business
  - 2300 Identify alternative strategies
  - 2400 Perform strategy evaluation
  - 2500 Document/present strategy
  - 2600 Strategy review
- 3000 Implementation approach
  - 3100 Evaluate the vision of the organization
  - 3200 Assess supplier alliances and relations
  - 3300 Assess current marketing initiatives
  - 3400 Approach to technology
  - 3500 Perform comparative analysis
  - 3600 Define approach
  - 3700 Review approach
- 4000 Identify e-business processes
  - 4100 Define core processes to support e-business
  - 4200 Group processes to include related processes
  - 4300 Create comparison tables for processes
  - 4400 Identify finalists
  - 4500 Evaluate finalist processes
  - 4600 Make final selection
- 5000 Competitive and marketplace assessment
  - 5100 Define on-going competitive assessment approach
  - 5200 Identify internal resources to participate
  - 5300 Investigate benchmarking
  - 5400 Identify specific sources of information
  - 5500 Define evaluation methods
  - 5600 Collect information
  - 5700 Organize information for long term use
  - 5800 Perform analysis
  - 5900 Present results of analysis

Figure 4.6 Sample project template for e-business implementation.

6000 Technology assessment
- 6100 Evaluate hardware in terms of e-business
  - 6110 Suitability and support of e-business
  - 6120 Identify missing hardware components
- 6200 Network assessment for e-business
  - 6210 Internal network capacity and performance
  - 6220 Security available and required
  - 6230 Extranet/intranet requirements
- 6300 System software assessment
  - 6310 Core operating systems
  - 6320 Data base management systems
  - 6330 Desktop systems
  - 6340 Utility software
- 6400 Test environment
  - 6410 Test hardware
  - 6420 Test network
  - 6430 Test software tools
- 6500 Development environment
  - 6510 Development hardware
  - 6520 Development network
  - 6530 Development software tools, languages, libraries, environment
- 6600 Identification of alternatives for e-business support
  - 6610 Hardware
  - 6620 Operating systems
  - 6630 Network software/management/security
  - 6640 Development environment
  - 6650 Test environment
- 6700 Define technology direction for e-business
  - 6710 Hardware
  - 6720 Operating systems
  - 6730 Network software/management/security
  - 6740 Development environment
  - 6750 Test environment
  - 6760 Interfaces with legacy and existing systems
  - 6770 New software
- 6800 IT staffing
- 6900 Develop comparative tables
- 6A00 Documentation of technology assessment

7000 Gather information on current processes
- 7100 Direct observation of processes
- 7200 Identification of issues in processes
- 7300 Interdepartmental interfaces
- 7400 Review of process documentation
- 7500 Assess current web activities

Figure 4.6 *(Continued)*

- 7600 Perform analysis and develop comparative tables
    - 7700 Determine fit with e-business
    - 7800 Documentation
    - 7900 Review current processes
- 8000 Define the new e-business processes
    - 8100 Generate alternatives for new processes
    - 8200 Assess alternatives in terms of regular/e-business
    - 8300 Develop comparative tables
    - 8400 Technology requirements for new processes
    - 8500 Staffing requirements for new processes
    - 8600 Compare new with current processes
    - 8700 Documentation
    - 8800 Review new processes
        - 8810 Current business
        - 8820 e-business
- 9000 Measurement
    - 9100 Identify areas of risk
    - 9200 Define e-business measurement approach
    - 9300 Infrastructure/technology/support
    - 9400 Measurement of current business
    - 9500 Measurement of web business
    - 9600 Measurement of web visitors
- 10000 Develop the implementation strategy
    - 10100 Define alternative strategies for processes
    - 10200 Define alternative strategies for technology
    - 10300 Define alternative strategies for organization/policies
    - 10400 Define alternatives for marketing
    - 10500 Conduct assessment of alternatives
    - 10600 Develop overall implementation strategy
    - 10700 Define prototype/pilot activity
    - 10800 Define phases for implementation
    - 10900 Review implementation strategy
- 11000 Define the implementation plan
    - 11100 Define implementation plan template
    - 11200 Identify specific implementation issues
    - 11300 Assess the project management process
    - 11400 Determine implementation leaders/team composition
    - 11500 Develop detailed plan and subprojects
    - 11600 Analyze the completed plan
- 12000 Implementation
    - 12100 Hardware setup for e-business
    - 12200 Network setup and testing for e-business
    - 12300 Firewall/extranet/security
    - 12400 Development environment setup

Figure 4.6 *(Continued)*

12500 Test environment setup
- 12600 Establishment of development standards
- 12700 Setup of quality assurance
- 12800 Installation of e-business software packages
- 12900 Establishment of external links
- 12A00 Testing of e-business software for production
- 12B00 Changes to current application software
- 12C00 Interfaces between current and e-business software
- 12D00 Implementation of marketing changes
- 12E00 Marketing campaigns
- 12F00 Setup of web content
- 12G00 Software development
- 12H00 Integration and testing
- 12I00 Quality assurance and integrated testing
- 12J00 Procedures and training materials
- 12K00 Operations procedures
- 12L00 Network procedures
- 12M00 Changes to current processes and workflow
- 12N00 Address current customers and suppliers
- 12O00 Conduct test of workflow and processes

13000 Post implementation assessment
- 13100 Gather lessons learned
- 13200 Identify unresolved issues
- 13300 Conduct usability assessment
- 13400 Conduct performance evaluation
- 13500 Assess impact on current processes
- 13600 Assess customer-supplier relationships
- 13700 Perform cost benefit analysis
- 13800 Define recommendations for later work
- 13900 Conduct review

Figure 4.6 *(Continued)*

When you make the presentations to management of the work here, keep these concerns in mind. Use the charts and tables that have been presented in this and previous two chapters. Remember that your political purpose is to gain support and enthusiasm for the improvement effort. Here is another tip. Show management the final two groupings of processes and the list of processes considered. In that way, you are allowing the managers to participate in the analysis. They are more likely to become involved and committed.

What are you asking for? It is simply the validation of the processes. A more

Element of the Scorecard	Score	Comments
Attitude of management before and after selection		
Range of processes		
Response of employees to process group selection		
Potential for Quick wins		
Awareness of process issues		

Figure 4.7 Process selection scorecard.

detailed analysis will follow. You are not asking for either vast amounts of money or personnel resources to implement change. These will come later. It is important that you segment the three issues: getting approval in concept, getting funding and money; getting resources. If you ask for all three at the start, you are less likely to get a positive decision.

EXAMPLES

ASC Manufacturing

ASC considered three process groups related to scheduling of work, allocation of resources, review of work, financial planning and analysis, project management, and issue tracking and resolution. In the end, the processes were scheduling, project management, and resource allocation. The method followed the one presented in this chapter. There was extensive input from managers, supervisors, and employees for the selection. This built up more support later.

Kosal Bank

Kosal Bank management realized that a great deal needed to be improved. Management awareness was not the problem. The issue was where to start. In the end, Kosal decided to focus on installment lending. This was a good choice because installment lending was of moderate complexity and was typical of the rest of the bank's operations, which would be improved later. The improvement methods could be demonstrated through installment lending and then applied and cloned for credit card and other lines of business.

Hetsun Retailing

In Hetsun action teams were formed at the start of the improvement effort. The teams first identified processes in each area. Next they formed alternative process groups along the lines discussed. They then proceeded to identify more than 230 issues and opportunities for change and improvement. Some of these were Quick Wins; others were for longer-term improvement. The issues and opportunities were then associated with the processes. The opportunities were also scored against the business mission and business issues. These factors were listed on improvement tables. Then the action teams voted on the opportunities and processes. These results were then reviewed by both the steering committee and the executive committee. This led to not only the process selection, but also to defining the Quick Wins.

For Hetsun, a large number of processes were selected. This was because many changes had to be implemented so that the company could get ready for e-business. However, the work was simplified since only a few transactions were selected for each process.

This application shows the flexibility of the method and approach in the book. You can move steps forward and accelerate the pace. Hetsun was driven by financial and competitive pressure to implement improvements to processes.

Lansing County

While there were many processes to consider, an earlier effort had centered on accounting. A general ledger system had been implemented and processes changed. However, it was clear that this was not a good choice, because it resulted in very limited benefits. Everyone was aware of this. Therefore, the choice of which processes to improve received a great deal of attention from management at all levels. Management decided to focus on operations where the bulk of the employees worked and where there had been little previous effort at improvement. People had just taken these processes for granted. Now it was their turn for change.

Management realized that if the improvement effort was initiated and handled by central administration, there would be problems and resistance from lower-level employees. Thus, a major effort was made to involve employees at remote locations in the improvement effort. It was also realized that each remote location had developed its own style and approach to the work. The new processes would have to result in more uniformity.

Simco International Bank

Simco is a new example that we introduce in this chapter. It will appear now and then in the book. Simco wanted to implement credit card services in the Southeast Asian region. There was much potential and also a great deal of competition. Simco had observed what other foreign banks had attempted and watched both failure and success. It was clear that most banks did not take into account the cultural and political factors in each country. Who controls the money, who makes decisions, and many other attitudes varied in the region. Simco management considered several approaches. One was the centralized approach to implementation and improvement. This was a failed model in general. Another one was to decentralize the effort totally. While this would have addressed cultural issues, it would have resulted in many differences and inconsistencies. The third approach is to implement a collaborative method across the region. This is the method that was chosen.

Action teams were formed for each credit card area: marketing and sales, application processing, payments, servicing, customer service, collections, and management reporting. Each action team crossed the region with one to two members from each country on each team. This required more coordination, so the bank implemented groupware and Internet collaboration tools such as teleconferencing. The focus was on a handful of transactions in each area. The implementation and improvement to current business was a success. Not only did Simco reach the schedule and benefit targets, but the effort instilled a new culture of collaboration that has persisted to this day.

LESSONS LEARNED

- Always focus on the critical processes. Do not be drawn off into supporting processes even though they are attractive.
- Concentrate on processes that cross organizations and are not within one department. Within a department, improvements tend to be fewer and have less benefit. Taking on processes across departments not only gives you more benefits, but also is politically easier to do since you are not subject to the whim of the management of one department.
- Focus on processes that are stable. Long-term existence without change is a good sign of great opportunities. Stability helps in several ways. First, you are not working with a moving target, so any improvement will more likely be lasting. Second, many exceptions and workarounds probably have been generated. By concentrating on the core transactions, you can begin to eliminate some of these. The downside is that because the process has been in place for a long time, there may be resistance to

change. On the other hand, there may be an attitude on the part of many that things need to change.
- Focus on processes for which improvements can yield a tangible economic benefit. Managers and employees have seen enough of fuzzy benefits. If you justify improvements on intangible benefits, you are just asking for trouble.

PROBLEMS YOU MIGHT ENCOUNTER

You are progressing further into process improvement. The steps in the preceding two chapters are fairly passive. When you select the processes for improvement, people start realizing that the situation is getting more serious.

- Employees are starting to become edgy about the change. They worry about their jobs and roles with the new processes. Try to involve some of the employees in the selected group of processes in thinking about potential improvements. You will do this in more depth later.
- Managers may want to derail your improvement effort here—before it starts. They might question the selection of the process group and indicate that the work in the group to do major improvements is just too much for the departments to handle. They may seek to defer the work. Your response is to indicate that only selected transactions will be considered—not everything. Next you will be indicating that employees will not be pulled from their jobs full time for any major length of time. You realize the importance of keeping the processes functioning.

WHAT TO DO NEXT

1. Write down several critical processes. For each, identify other processes that should be grouped with it based on the discussion in the chapter. You can now prepare a table in which the columns are the related processes, the reasons, and the association.
2. You can begin to determine how the organizations and departments are involved in the processes within a group. You can prepare a table in which the rows are the departments and the columns are the processes in the group. The table entry is the degree to which the department is involved in the process.
3. Consider the elements of the mission or vision that were discussed in Chapter 2. These can be rows in a new improvement table. The processes in a group are the column headings. The entry is the benefit and support of the mission or vision element if improvements are made to that specific process.

CHAPTER 5

Examine Your Current Business Processes

INTRODUCTION

Remember that the chapter headings are not numbered as steps. This is done to encourage you to think of doing parallel work. So you might begin collecting information on processes before the process group is approved for the work.

Before we start you should keep in mind some basic truths about process improvement:

- Process improvement is political. This has been said a number of times. Here you must be aware of how you position process improvement to the employees, customers, or suppliers. If you don't think about this or do it wrong, the entire improvement effort can be placed in jeopardy. Thus, we will give this special attention.
- Collecting information is the principal means of interacting with employees and supervisors involved in the work. This is why data collection is so important. It must be planned.

You will be gathering all types of information, including the following:

- An understanding of how the transactions work
- Identification of potential problems and issues in doing the work and their impact on the processes
- Identification of shadow systems and their roles in the processes
- Awareness of how different processes interrelate through the business processes

- An understanding of the steps necessary to get people on board the process improvement effort
- Identification of individuals who may potentially resist improvement
- Identification of individuals who are in favor of change
- Identification of people who have de facto power through their knowledge and seniority with respect to the business process
- Identification of people who have extensive business knowledge
- Identification of workarounds and exceptions

As you can see, this is a lot to do in a limited amount of time. You really will not achieve these goals if you rely on interviews.

OBJECTIVES

TECHNICAL OBJECTIVES

The technical objective is to gather the information in the previous list. Another purpose is to understand the work sufficiently to be able to propose new ways of doing the work and to uncover Quick Wins that lead to longer-term improvements. You do not have the luxury of going back to the department constantly. By using the action team approach, you will get more involvement from the employees. Here are some more specific technical objectives:

- Determine which transactions will be considered for process improvement and Quick Wins.
- Define issues and opportunities that are affecting the work at the transaction level.
- Identify potential improvements and develop an initial assessment of their impact and ease of implementation.
- Validate all findings with supervisors and employees.

BUSINESS OBJECTIVES

The business objective is to not only understand what is going on, but also to get the people to understand the need for change. While this is also political, it is the key business objective. The basic point is that the only way people are willing to change is if they see the problems and shortcomings in what they are doing. This is exactly what happens in drug and alcohol rehabilitation. A person must admit that he or she has a problem in order to improve.

Political Objectives

There are a number of political objectives here:

- You want to line up support for change. You want to get people involved in the improvements. You want them to own the improvements. Taking ownership is a key goal since you are seeking not only improvement, but lasting improvement.
- You want to identify who might be resistant to change and to work with them to understand their concerns. If you try to overwhelm them by appealing to management, you will just engender latent hostility.
- You are trying to create a sense of excitement and enthusiasm. You are raising hopes for change. You will then show those involved that they have an important role to play in making the changes come true and remain.

END PRODUCTS

There are a number of end products, including the following:

- Problems and issues that are being faced in the processes.
- Understanding how transactions work and how they interface with each other.
- Defining the boundaries of process improvement. Here you are defining which transactions will be addressed and what exceptions and other work will not be part of the effort.
- Increased involvement of the action teams and other employees.
- Suggestions by the employees about potential changes and improvements.
- Detailed definitions of the transactions, including the business rules.
- Identification of constraints due to organization structure, policies, and procedures.
- Understanding of the informal power structure around the business processes.

METHODS

Where to Start

There are several basic ways to collect information. The first is from passive sources. Here you will examine files, other reports, previous efforts, and written memos and letters. This will give you some background. Read this material actively. Try to answer the following questions:

- Who are the people suggesting improvements and change? Who is raising arguments against change?
- What efforts have been tried in the departments in the past? Note that you are at the department level. What has happened with this work?
- Has there been any effort to deal with exceptions and workarounds?
- What measurements of processes have been done?
- What problems and issues have been raised recently? Maybe, you can address some of these in the improvement effort.

Four other methods of gathering information are interviewing, observing, getting trained to do the work, and having informal conversations. We definitely prefer the last two. Interviewing has many drawbacks, including these:

- A person being interviewed will tell you what is on her or his mind as of that moment. There is little structure.
- You receive opinions without evidence.
- Most people have not been trained to be good interviewers in terms of the sequence of questions, follow-up, observations of body language, what tone of voice means, and so on.
- You are likely to get canned answers to questions.

We are not saying "Don't interview." Rather, you should not expect much from interviews and you should not overly rely on them.

Observing work and transactions is an excellent means to gather data. It is a firsthand approach. However, don't think that by observing what is going on, you will catch everything. You must go back to the work and observe it again and again. By observing different people at different times, you will understand the work much better. In several cases, including ASC Manufacturing, we had to visit each shift and observe the work process throughout all three shifts. A lot goes on during the third shift. You don't want to miss it.

A third technique is to be trained in the process. This is excellent. You can get the following benefits from being trained.

- You can learn how to do the work.
- You will learn some of the business rules and terminology.
- You can detect exceptions and workarounds.
- You can judge how effectively people are using computer systems.
- You can assess the quality of the procedures and the training.
- You can observe the work.

The fourth technique is to engage employees during lunchtime or breaks and discuss things. We suggest that you go out to the location where people smoke.

One of us smokes a pipe. Smoking with the people or being with them tends to make them open up more to you.

You want to maintain files of what you are doing. Here are some areas in which to organize your information:

- List of contacts and their roles
- Log of documents that you are collecting
- List of issues and opportunities
- List of lessons learned and experience
- Potential improvements and their benefits and impacts if not done
- Implementation considerations
- Resources needed for improvements

In terms of internal documents, here is a list to get you started:

- Organization charts. Try to get several years' worth of them. They can tell you who is rising and falling in the organization.
- Department budgets.
- Annual reports to management from the departments.
- Procedures and forms used in the departments that relate to the work.
- Internal memos and notes from department staff.
- Training materials used in the departments.

Perform an Initial Assessment

Now you are ready to get out there. You will first conduct limited interviews from the top down. These should be very short and act as a political introduction of the improvement effort. You will be telling the staff members about the work. You will want to get from them answers to the following questions:

- What are the goals of the processes?
- What problems and opportunities do they know about for the processes?
- How has the process changed over the past year?
- What is their attitude toward the work?
- Who do they suggest you contact?

Now you have worked your way down to supervisors. Have them take you on a tour of the work and process. Get an agreement on how to collect information. Push for being trained and for direct observation. They will be concerned that you will get in the way and impede work. You can head this off by indicating that you are sensitive to this and that when it is busy, you will make few, if any, demands on their time.

Examine Specific Transactions

This section pertains to considering an individual transaction. Here are some issues to address:

- What are the specific steps in the transaction?
- What are the volume and frequency of the transaction?
- Are their exception conditions? How do you know an exception when you see one?
- What workarounds are used? Why are they used?
- What shadow systems are in place? Why were they created? What condition are they in?
- What is the boundary of the transaction between departments? How does the hand-off work for transfer between departments?
- How are rework and errors addressed?
- What statistics are kept in terms of the work and transaction?

You also have the opportunity to pose some other interesting questions. Here are some that we have employed repeatedly:

- What is the most interesting work that you do?
- How did you learn to do the work?
- If you change anything, what would you change?
- If you had money to do something, what would you do?
- If you had recurring money, what would you do?
- What problems do you have in doing the work?
- What is the most difficult transaction that you do?
- If you were to train someone new, how would you do it?
- What could be done to make the work easier?
- What are the problems with the computer systems?
- When do peaks of work occur? What do you do differently at such times?

It is just amazing what people answer. In the implementation of a drugstore system, we asked these questions of the store manager. She responded to the question as to what she needed with, "I want another pair of handcuffs." We were, to say the least, taken aback. Why did she want this? She indicated that shoplifters come into the store in pairs. She only had one pair of handcuffs. She also said that she had requested this many times, but was ignored. This was a Quick Win that had little to do with computers or theory of processes. It was a very big hit. After that, there was only support for process improvement because we got *something done fast.*

Here is another example. At ASC Manufacturing on the third shift, the area where people ate their meals was locked up for security reasons. People had to eat on the floor. There were no tables and chairs. Like the previous case, they had requested these things, but nothing happened. We got them tables and chairs and in return received loyal support for change.

Now you will also be collecting forms and logs and other similar materials. Here are some things to look out for:

- What is the date of the form? Is there a number? The date and number indicate that it is formal form. This means that at some time in the past the process was analyzed. If it is undated, how long has it been used?
- Look at the layout of the form. Does it appear well organized? Forms design used to receive a great deal of attention. Now it is less so.
- Ask for several completed forms. See if people are writing down things outside of the information required by the form. How does the form hold up under use?
- Are there multiple copies of the form? Where do they go? How are they used?
- Are there instructions for completing the form? Try to follow these. Are they clear?

Changes to frequently used forms and eliminating or combining forms are candidates for Quick Wins. You can help people's productivity by improving these forms.

Now turn to the logs. Why is the log being kept? Is it for defensive reasons? An example might be that the transactions are logged before they are sent to the next department. Was this done to cover one's trail if there was a later problem?

You will be having the action teams identify problems and opportunities as discussed in Chapter 2. Here you will have them write down for each opportunity the following:

- Date
- Title of the opportunity
- Name of person
- Process involved
- Transaction involved
- Problem or opportunity description
- Impact if it is not addressed
- Benefits to implementing or fixing it
- Implementation suggestions and effort

As you are collecting the information, you will want to analyze it. Improvement tables and the scorecard are addressed later. Here are some more detailed suggestions:

- Write notes or e-mails to supervisors and managers thanking them for their support. Do this on a regular basis. Don't take the support for granted.

Transaction: ____________________

Step	Who Does it	What is done	Issues	Impact of issue

Figure 5.1 Analysis table for a current transaction.

- Enter the issues and opportunities in the issues database. Do the same with lessons learned from experience.
- Develop some transaction analysis tables (discussed later).
- Make notes of ideas submitted by employees and make sure to write down the employee's name and the date that each idea was provided. You will want to give the employee credit later.
- Organize the opportunities that are submitted by the action teams. Review these for understanding and completeness.

For each current major transaction, prepare the analysis table shown in Figure 5.1. Note that the first three columns are playscript and are 2,500 years old—dating from the Greek plays. This is a proven technique and better than complex diagrams.

This table can help identify very specific issues and impacts, which is very useful when you later propose specific process changes and Quick Wins.

You can carry this a step further and create another table (Figure 5.2).

You can now roll up the transactions and get a summary table. An example appears in Figure 5.3. You will use the summary table a number of times.

As you are developing these tables, have the employees review them. Here is a tip. Make some intentional mistakes that they will detect. They will tell you how to fix them. This will get them more actively involved in reviewing the current work. We have found that using these tables is much more user-friendly and elicits more feedback than complex diagrams or massive amounts of text.

Transaction: ____________________

Step	Guidelines for doing the work	Errors	How rework is done	How problems are detected

Figure 5.2 Additional analysis of a transaction.

Process: ______________________

Transaction	Issues	Frequency	Volume	Errors	Rework

Figure 5.3 Summary analysis table for the current transactions.

You can also employ the spider chart method that was used in previous chapters. Here you might choose the following dimensions:

- *Work.* Amount of work involved in the transaction (frequency, volume, etc.)
- *Deterioration.* Deterioration of the transaction over time
- *IT issues.* Degree to which IT issues are affecting the transaction
- *Issues.* An assessment of process issues and degree of impact
- *Exceptions.* Extent of exception work in the transaction
- *Shadow systems.* Role and importance of shadow systems in doing the transaction
- *Benefit.* Potential benefit and impact of improvements
- *Impact.* Potential impact if no action is taken

You can now compare two transactions, as shown in Figure 5.4. These are for Kosul Bank. One deals with initially calling customers when they are behind in payments (labeled A) and the other is for charging off a loan as a bad debt (labeled B). Of course, these are subjective. However, they will be useful in getting

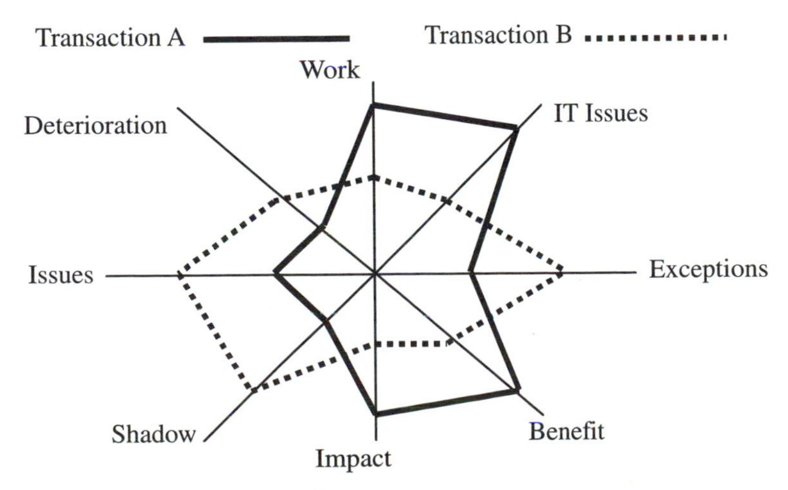

Figure 5.4 Example of chart comparing two transactions.

people to understand the different characteristics of specific transactions. You can develop a similar chart to compare processes as well.

It is useful to make some observations about Figure 5.4:

- The first transaction is of much higher volume, impact, and benefit than the second. Charging off a loan is a slow, tedious, and individual process.
- The second transaction has suffered more in issues and deterioration, but this is outweighed by the other factors favoring the first transaction.
- The first transaction has IT issues that can be addressed. Some of these could be Quick Wins.

You can use this method to select individual transactions.

Example of Analysis: Kosal Bank

Kosal Bank employed the previous approach to analyze its transactions. Space does not permit us to present all of the transactions. However, two of these that were quite different can be discussed. Both occurred in collections for installment lending. In collections, the first step is to call any customer who is X days delinquent—that is, payment for the loan should have been received X days ago, but was not. When you reach the customer, you play a prerecorded message. The second transaction consists of the old combined manual and automated method for calling customers who are more than 30 days overdue. These transactions were changed through process improvement. The new ones will be discussed in Chapter 7.

The analyses of the transactions are given in Figures 5.5 and 5.6, respectively. In the first example, the computer system queues and displays the accounts. The employee reads the number and dials the customer. Once contacted, the employee greets the customer in English. Upon acknowledgement, a prerecorded message reminds the customer that payment is overdue. If the customer responds and wants attention, the employee must route the call to a collector. This is a high-volume operation, but it is very short. Also, it is not as critical as the actual collections activity in the second figure.

In the second example, a computer system issues a batch report. This is the installment lending system. The collector finds the most delinquent account and then looks for the log of previous calls to the customer. The collector reviews this material. Then a call is made to the customer. This is the action. You can reach the customer or you can get no answer, a busy signal, and so forth. The result of the call is the customer response, if reached. The most desirable response is a promise to pay. The collector records these results and any comments on the card and then goes on to the next customer.

While this is not a recent example, this one was chosen because it reveals some of the problems with computer systems, manual procedures, and policies. The examples will be used in later chapters.

Step	Who does it	What is done	Issues	Impact of issue
1	Computer system	Queues calls		
2	Employee	Calls customer	Manual effort	Lost productivity
3	Employee	Gives greeting	Inconsistent message	
4	Employee	Plays recorded message	Only in English	Some customers cannot understand
5	Customer	Responds and wants attention	Has to wait	
6	Employee	Manually routes call to collector	Customer waits longer; work is disrupted for the collector	Poor service; reduced productivity

Figure 5.5 Kosal Bank's initial collection call example.

Assess Exceptions, Workarounds, and Shadow Systems

You have identified instances of exceptions, workarounds, and shadow systems. Here are some questions to address for these issues:

- What is the frequency of exceptions? Should any be considered in process improvement?

Step	Who does it	What is done	Issues	Impact of issue
1	Computer system	Produces batch report for collectors		
2	Collector	Reviews report and goes to manual comment file	Very labor intensive	Low productivity
3	Collector	Makes call to customer		
4	Collector	Converses with customer	Inconsistent training	
5	Collector	Manually records action taken and results obtained	Manual, inconsistent	Lack of controls; low productivity

Assess Exceptions, Workarounds, and Shadow Systems (H3)

Figure 5.6 Kosal Bank's subsequent collection call example.

- For each workaround, what caused it to be created? What are the underlying problems?
- For each shadow system that you found, what caused it to be established? How is it used in regular work and for exceptions?

The purpose here is for you to determine what part of this is going to have to be included in the process improvement. Remember that the default is not to consider a particular exception. If you are asked if an exception is to be included, you can answer, "Probably not unless it occurs frequently and affects a substantial part of the work."

Develop Improvement Tables and Process Scorecards

Here are some of the improvement tables that we use for a transaction:

- *Process steps versus process steps.* Here you are assessing similarities among transactions to make the future improvement work easier and more consistent. The entry is the degree to which the step is the same in two transactions. You might want to enter text here to show how they are similar.
- *Process steps versus infrastructure.* How do the computer system, network, and other parts of the infrastructure support each process step? You might enter comments on impacts and support. This will help you see which steps are not supported as well.
- *Process steps versus potential technology.* Use this table to uncover opportunities for new systems and technology.
- *Process steps versus process issues.* This table shows which steps each issue addresses.
- *Process steps versus organization*. This table indicates which organizations perform what transaction steps. It is useful in showing inefficiency and complexity in transactions.

Now you can move to the transactions with the use of the following tables:

- *Transactions versus organizations.* This table reveals how organizations are involved in the work. It also shows you which departments are the most heavily involved in the work.
- *Transactions versus performance criteria*. This table lists frequency, volume, and other attributes. These are performance criteria. The entry is a numerical measure. The table allows you to compare the relative importance of the transactions at the level of the department.
- *Transactions versus business issues.* In Chapter 2 you identified key business issues. Now you can put this information to use and create a table that shows which issues are most relevant to the specific transactions.

- *Transactions versus infrastructure.* The elements of the infrastructure appear as columns. The entry is the impact or problem that the infrastructure element causes the transaction. This table can point the way to which infrastructure elements need to be changed.
- *Transactions versus process issues.* The common process issues are the columns. The table entry is the impact of the issue on the transaction.
- *Transactions versus potential technologies and systems.* This table reveals the potential benefit of a specific new technology or system on a transaction.
- *Transactions versus IT issues.* This table can be used to indicate the impact of known IT problems on the work. As such, it can help build support to address the issues.

You can move further up to the process group and create three more improvement tables:

- *Process group versus business objectives.* This table indicates the degree to which the current processes support the business objectives.
- *Process group versus business issues.* This table indicates the extent to which the current processes contribute to the business issues.
- *Process group versus current IT architecture.* The table reveals the degree to which specific architectural components impact individual processes in the group.

There are two kinds of scorecards for you here. The first is the scorecard for the business process. The second is for you to assess how you are doing. These are given in Figures 5.7 and 5.8, respectively. Some remarks about the first scorecard are in order. First, you will want to define your own factors. These are just suggestions. Second, this scorecard is useful for showing management where things stand prior to improvement. You will be revisiting this scorecard later as you implement Quick Wins and longer-term improvements. Third, you will want to involve the action teams in completing the scorecard.

GAIN SUPPORT FOR CHANGE

In analyzing the processes, you are selecting specific transactions and examining them. This is the technical side. It is at this stage of the improvement effort that you must achieve the following goal.

> Employees and supervisors acknowledge the problems in the work and transactions, and they support changes and improvements.

If you don't get this now, then much of your work later is going to be for nothing. Why? Because the employees do not see the need for change. Then when you

Factor	Score	Comments
Number of people involved		
Turnover of business staff		
Number of exceptions		
Number of shadow systems		
Quality of training materials		
Quality of procedures		
Extent of current automation		
Existence of measurements		
Cost of doing the process		
Impact of leaving the process alone		
Number and severity of issues		
Impact of systems and technology on the process		
Extent of potential opportunities		

Figure 5.7 Scorecard for a business process.

have implemented the change and are patting yourself and the team on the back, deterioration begins. The new processes will potentially develop gangrene—not a pretty picture. That is why you must involve many different employees and supervisors in this work. They are not involved full time. As you saw from the previous discussion, their involvement is limited.

Element of the Scorecard	Score	Comments
Extent of suggestions provided by employees		
Participation quality and volume by Action team members		
Support by supervisors and employees		
Extent of information available on the process		
Involvement by managers and employees in reviewing the work		
Suggestions for Quick Wins and improvement		

Gain Support for Change (H3)

Figure 5.8 Scorecard for the improvement work.

EXAMPLES

ASC Manufacturing

At ASC, a major effort was made to involve many employees. Of the more than 2,000 people, more than 250 were contacted and either interviewed or worked with through conversations. This led to grass roots support for change.

Kosal Bank

There was little need to get support for change. The turnover in some units in the bank was over 30% per year. The challenge was to understand the problems and opportunities. The employees felt overwhelmed by the many problems involving policies, procedures, systems, and infrastructure.

Hetsun Retailing

At Hetsun there was only a core of 50 or so people who really knew how the processes worked in different departments. This made the analysis easier. However, efforts will still made to validate the findings and results with many other employees. Getting the lower-level employees on board proved very useful. Some became very valuable members of the action teams. They came up with some of the best Quick Wins.

Lansing County

Lansing is a bureaucracy. People had been doing things the same way for many years. It was not that there was resistance. Many people had given up hope for improvement. Thus, it was important to identify Quick Wins and to implement some changes quickly. So an effort was made to make some changes without waiting for the complete definition of the new processes. Changes were selected based not only on impact but on stability after the processes were changed later.

LESSONS LEARNED

- When collecting the data and making your observations, dress like the people you will be working with. Do not take too many notes. People then get more defensive. Try to use the same words and terminology that they use. That will not only get you closer to them, it will also reinforce your understanding of the words.

- Ask people where they spend their time. This will show you indirectly where there are problems.
- Stress simplification and minor changes at the start. Avoid anything about job changes or elimination.
- Find some interesting examples of transactions that are humorous. Use these with management. It shows that you are on top of the work.
- Move around departments rather than staying within one department. You want to show up in different departments. In that way, you will appear unbiased.
- If people resist being involved, keep them informed. This will help you wear down resistance.
- When someone suggests a change, encourage the person to discuss it. Ask for the person's opinion on why the change was not implemented before.
- Perform your analysis work in the departments near the work as much as possible. You will not only pick up more information, but you will also show your interest in the work done in that department.
- Try to identify how a process has changed over time. You are trying to detect signs of deterioration. Ask employees how they did the work the previous year and how it has changed.

PROBLEMS YOU MIGHT ENCOUNTER

- Some people resist cooperating by indicating that they provided this information before. Perform your homework prior to getting out to departments. Find out what information was gathered and review it.
- Other projects were done and nothing happened. People are then naturally inclined to view the improvement effort as another in a chain of failures. What is your response? Indicate that this is different because more people in departments are getting involved in the work. Also, point out that there is an emphasis on Quick Wins as well as long-term improvement. Then ask them what problems they have.
- Some may point out that the process is already dependent on systems and that these cannot be easily changed. Therefore, there is little point in trying to improve such processes. Point out that you are looking at a group of processes so that there is much more involved.

WHAT TO DO NEXT

The best thing to do is to get some hands-on experience and see what happens. You should go out and observe a specific process in your organization. Take one that is not too complex, and identify two different transactions. Go through the actions that have been defined in this chapter. This will get you started on the road to being proficient at process improvement.

CHAPTER 6

Develop a Process Plan

INTRODUCTION

There are business plans, department or division plans, and information technology (IT) plans? These all address organizational concerns and needs. None of these specifically deals with where the money is either won or lost. After all, the success or failure of an organization is determined by its business processes. That is why it is interesting that so little attention has been devoted to this area—developing the process plan.

This chapter details how to go about developing and using a process plan. This is not something you do for all processes. You will initially want to do it for the key processes in your process group for the improvement effort. You can then later expand it to, probably, no more than 10 processes.

What is a process plan? It is the counterpart of the business plan. The process plan includes the following information:

- How the process works and relates to other processes
- The process scorecard, as developed in Chapter 5
- The challenges and issues that relate to the business process
- The impacts of the process on constituent audiences such as customers, employees, management, and suppliers
- The long-term vision of the process—the benefits if you had the perfect process
- The role of automation and IT in the process over the long term

Why bother with this? Don't we have enough to do? Well, you can go right past this chapter and start designing the new process. However, you are less likely to get management support since the long-term goals of what you are trying to achieve are not clear. There is another reason to create the process plan. The process plan supports long-term stability of the business process and change. Without it there is less stability. People will tend to introduce unplanned changes.

Here are some reasons why the process plans are not developed.

- There is no widespread recognition that this is an accepted step of work. Business planning and IT planning are accepted widely, but process planning is not as widely recognized, even though successful firms credit part of their success to having a plan and vision for their processes. In the standard Six Sigma application, the focus is on specific improvement steps and less on the long-term vision of a process.
- People are organization focused. That is, if you fix or improve the organization structure, then the process improvements will follow. The reasoning is that if you get good people, good processes will follow.
- Some managers pay lip service to the importance of processes, but give them little attention. There is an assumption that the employees will handle the process. This is evident in firms where there is little training in the process except on-the-job training. Individuals are often hired into the departments based on previous work that provided them with similar experiences. Managers say, "Why train them if they have done it before?" This can be a big mistake. One company that had a customer call center with three shifts hired a manager from a different company for each shift. Each implemented the practices that he or she had followed at a previous job. They were all different in major ways, causing many process problems.
- Some managers view processes as simple in that after you fix them, they will continue to run for a long time without major attention. However, as was discussed earlier, deterioration in a process can and does set in fairly quickly after changes.

Some comments about process plans are useful here. The development of the process plan can be carried out in parallel with other work so that there is not much of an impact on the overall schedule. Another point is that the process plan indicates where the process should eventually end up. It may be a long road with many turns between where you are now and the eventual goal. It may not, for example, be possible to attain the goal in a few years because there are organizational, regulatory, or technology barriers. Developing the process plan helps you to understand their barriers. Thus, the improved transactions that you will define in the next chapter may only be an intermediate step toward the eventual goal. By having the process plan, you will know after improvements have been made how far you still have to go.

OBJECTIVES

Technical Objectives

The technical objective is to develop process plans in a collaborative way. By creating the process plan, the managers and employees become more aware of the impacts of the problems in the current work and processes. They can also see the effects of constraints and factors that cannot be changed, but which have a severe impact on how the process work is performed.

Business Objectives

The business objective is to provide employees and managers with a vision of what the future process might be and look like. This then gives people a goal or target to aim for. It also helps produce more stable requirements. This can help them understand the impact of the problems in their current work.

Political Objectives

One political objective is to solidify support for process improvement. Think of it as a carrot and stick approach. The stick is the negative effect of the problems and issues in the process. This can get people depressed and lower morale—even if you have defined some Quick Wins. Even if you define the new processes, this is just a step on the road toward improvement. Hopefully, with success there will be many more improvements. However, what is important is that people can visualize what a greatly improved process would be like. They can then picture themselves in this process. The approach of developing the long-term vision is part of not only business practices, but also of religious, political, and sociological theory as well.

END PRODUCTS

The process plan includes the following elements:

Current Situation

- A description and analysis of how the process works and interfaces with other processes
- An overall assessment of the current process in terms of major problems and issues

- Impacts of the problems and issues on the organization, IT, employees, management, customers, and suppliers
- The process score card summary of the current process

Long-Term Vision

- A description of the long-term process
- Examples of transactions in the long-term process
- Comparison of the current and long-term transactions
- Benefits of the long-term process
- Potential process scorecard summary for the long-term process
- Barriers and constraints that prevent the short-term attainment of the vision
- Critical success factors in moving toward the long-term process

Note that the expression "long-term" is used instead of the word "future." This is intentional. The future process will be established in the next chapter. As such, it is usually just a stepping store toward the long-term process.

How are these end products and the process plan employed?

- They provide input into the development of the improvements to the work and processes. That is the tangible short-term benefit.
- The results can be used in the development of the business plans. Business plans provide targets for growth, revenue, sales, and so on. As such, they are detached from the actual work, being an abstraction. With the process plans in place, you can now try to relate the factors in the business plan with the process plan. After all, how the processes work is what determine the success of the business plan.
- The IT plan addresses objectives, issues, and provides a road map for IT. In addition to improving the IT architecture, IT work attempts to support the key business processes better through doing work in the plan. If you have process plans, IT knows better what is required for the longer term.

METHODS

Where to Start

In the previous chapter you analyzed the current work and performed substantial analysis. This is the natural place to start. Here are the steps that you can follow:

- Identify the constraints and barriers in the current processes and work that prevent improvement.

- Prepare graphs, charts, or tables that illustrate the impact of these constraints on the process.
- Define the future vision of the long-term process. This is almost impossible to do from the top down. Instead, you can proceed with specific transactions and move from the bottom up to the process.

Identify Constraints and Barriers

Now consider how some specific transactions work. Take a look at the following list of constraints.

- Policies
- Organizational structure and staffing
- External regulations and cultural factors
- Other surrounding processes
- Customers and suppliers
- Technology relied on by the processes
- Systems supporting the processes
- Management factors and politics among and within departments

You can also break these up into more detailed factors. You can employ in Figure 6.1 as a checklist.

Some comments are useful here on the impact of these factors on processes.

- Policies can be very restrictive to work. Policies typically were set in place years ago. No one questioned them; they just were there. If a policy is a real problem, the people doing the work will not attack the policy directly. They still want their jobs! Instead, they will create exceptions to get around the policies that don't work. This obviously creates many headaches. If, on the other hand, they follow the policies, additional problems can be created later. In some cases, policies create more work. Consider the collection examples of Kosul Bank. The original policy was to call customers three days after their payments were due. However, allowing for the mail and check processing, this time interval was too short. So a major problem was that many unnecessary calls had to be made to the customers.
- Organization structure can affect a process in different ways. First, the positions and the work may be mismatched. Job descriptions may not reflect work. Pay scales may not be correct for the level of complexity. These all point to morale and day-to-day work problems and issues. Usually, the organizational structure badly meshes with processes since the processes changed but the organization did not. On the other hand, management could have imposed a new organizational structure that does not fit the work.

External factors
- The culture of the country or region
- The distribution of the work across a region or internationally
- Regulations and rules imposed by the local and national government
- Geographic factors related to traffic, climate, etc.
- Availability of a workforce
- General economic conditions

Organization related factors
- Rules and structure imposed by a remote headquarters
- Level of staffing and their qualifications in the processes
- Internal politics

Technology and infrastructure-related factors
- Internal systems used directly in the processes
- Technology (network, hardware, and system software) that supports the processes
- IT goals and policies
- IT priorities
- IT workload
- Buildings and facilities

Policy and process factors
- General company policies
- Policies for the specific processes
- Policies for related processes
- Procedures and functions of other processes

Figure 6.1 List of potential constraints for business processes.

• There are many staffing issues. The work may be best done by people who are better trained and have more knowledge than those doing the work. This can create more errors and exceptions.

• External regulations and rules can really dampen or warp processes. For example, a country may require that specific tasks be done in a certain way and that various manual records be established and maintained. This can be a heavy burden on a process.

• Cultural factors play a role in processes. In some cultures, what the direct supervisor says is the final word. This creates inconsistency in work among different supervisors. Employees in other cultures may be shy and resist bringing up issues or questions to management. Then the errors multiply and expand.

• Processes are always interdependent. Let's say that you have selected process B to be part of the group of processes you are improving. Process A is in another part of the company. Work starts in A and goes to B. The potential impact is obvious. What is not evident is a downstream process, C, that depends on process B, the one you are improving. Work gets into C and can be returned or rejected because of politics or quality.

• Customers and suppliers play significant roles in affecting the processes. For example, you may want to implement e-business, but you find that the suppliers or customers are not ready. This can cause you to abandon the effort. Customers may also be in the habit of having work performed in a certain way. If you change the process, you could drive the customers away.

• The technology acts as a constraint because it limits both the capacity and capabilities that are possible to implement.

• More directly important to processes than technology are the application systems that directly support the processes. You have to assume that the application systems are constraints in most cases since you do not have the time to replace them quickly.

• Management factors are many and varied. There can be personality issues. There are issues with a manager's style. Political hatred and infighting can be persistent among departments. All of these factors and others can inhibit any significant process improvement effort.

You can now determine the relevance and degree of impact on the specific transactions. A spider chart here can be useful. Figure 6.2 presents two charts for the two examples of Kosal Bank that were presented in Chapter 5. Recall that the first was the initial call to a customer who had not paid, and the second was to contact the customer who had been delinquent in payments for some time.

For the initial call, customer and policy issues play major roles. People feel that they are being harassed. Many claim that they paid on their loans and should not be called so soon. The technology and systems are factors in the work as well.

For the standard collection contact there are significant effects from the systems and technology areas, other processes, and organization structure. The collections effort is hampered by a disjointed operation that is spread across many independent offices in the region.

Assess the Impact of the Constraints

Having figured out which constraints play major roles in affecting the transactions, let's identify the potential impacts of these constraints. Here is a list of six of the major impacts.

- *Cost of doing the work.* This is overall cost and not just the IT part. Here you would want to include labor, overhead, IT, and facilities costs, for example.
- *Elapsed time to perform a transaction.* Elapsed time is what counts in productivity and to the customer or supplier. You can have a transaction that takes a short time, but it must be split up into pieces due to poor organization or some other problem.

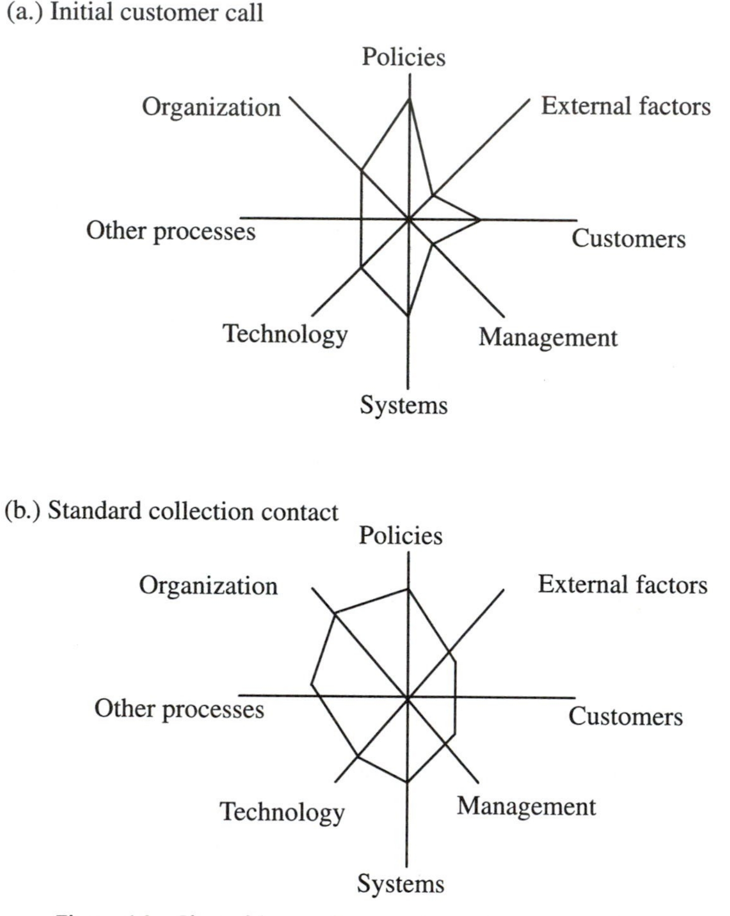

Figure 6.2 Chart of degree of impact of constraints on transactions.

- *Effort required to perform a transaction.* This is the labor needed to do the work. A badly working transaction will tend to necessitate higher-level and more trained staff as will complex transactions. In insurance, for example, the level of effort needed to give someone a quotation for an insurance policy is much less than that required for a claim.
- *Errors and rework created and traceable to the constraint.* If an application system experiences sporadic errors, then this creates errors in the data that have to be corrected. The same is true for work captured according to one set of business rules and policies that do not fit those of the next group handling the work.

- *Volume of work that can be performed.* This is the measure of capacity to do the work. Poorly laid out facilities or a faulty organization can impact the potential capacity.
- *Exceptions and workarounds necessitated by the constraint.* This one is famous or notorious, depending on your point of view. If the computer system does not complete the entire transaction, then the department may have to create shadow systems. If the policies are not complete, then exceptions are created.

You can develop a table of the constraints and their impacts. Figure 6.3 shows a table for the two transactions of Kosal Bank discussed in Chapter 5.

(a.) Initial customer call

Constraints	Impact	Comments
Policies	Severe	Excessive numbers of calls are generated
Organization	Minimal	Some improvement is possible at the department level
External factors	None	
Other processes	Minor	Could improve the loan accounting update
Customers	Severe	Major customer dissatisfaction
Technology	Moderate	Older technology in use
Systems	Moderate-severe	Systems are not integrated
Management	Minor	

(b.) Standard collection contact

Constraints	Impact	Comments
Policies	Moderate-severe	Policies are not standardized
Organization	Severe	Organization is distributed
External factors	Minor	
Other processes	Moderate	Follow-up for recurring delinquent customers is not good or well organized
Customers	Minor	
Technology	Severe	Lack of modern technology
Systems	Severe	Old systems
Management	Moderate	Lack of management data

Figure 6.3 Table of constraints and impacts.

Next, your analysis reveals that some constraints are more severe than others. There are several ways to show and examine this analysis. One is to use a table in which the rows are the constraints and the columns are the impacts noted in the preceding list. You can then insert a numerical entry of 1 to 5, where 1 is a minor impact and 5 is a severe impact. Figure 6.4 presents an example. Note that an "overall" row has been added at the bottom. This is a weighted average.

The alternative that we tend to use, since it is more graphic, is a spider chart. Figure 6.5 shows the corresponding example. From experience, you will to have both available.

(a.) Initial customer call

Constraints	Cost	Elapsed time	Effort	Errors	Volume	Exceptions
Policies	4	3	2	2	5	1
Organization	3	1	2	2	4	1
External factors	1	3	1	2	4	1
Other processes	1	1	2	1	2	1
Customers	4	2	2	3	5	1
Technology	4	4	4	3	4	2
Systems	5	5	4	3	4	2
Management	1	1	1	1	1	1
Overall	3	4	3	2	5	1

(b.) Standard collection contact

Constraints	Cost	Elapsed time	Effort	Errors	Volume	Exceptions
Policies	5	5	4	3	1	4
Organization	5	5	5	2	2	5
External factors	2	3	3	1	1	2
Other processes	3	3	2	2	1	3
Customers	2	4	3	2	2	2
Technology	5	5	5	3	5	5
Systems	4	5	5	3	5	5
Management	2	3	4	2	2	3
Overall	4	5	5	3	2	4

Figure 6.4 Scoring table of constraints and impacts.

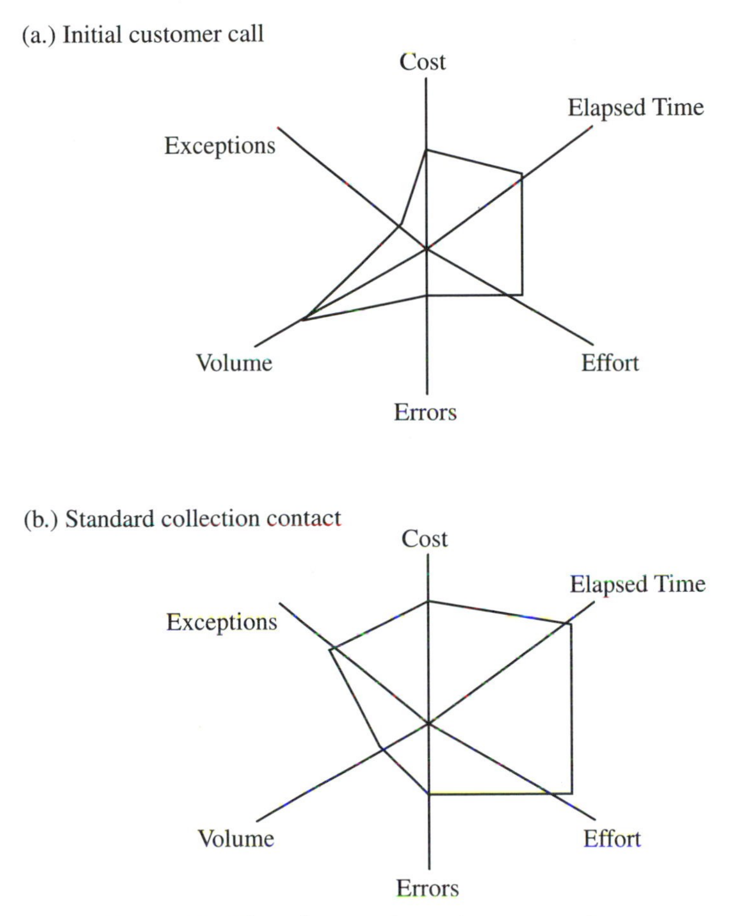

Figure 6.5 Chart of impact of constraints on transactions.

Determine Long-Term Process Transactions

So far you have considered which constraints have the most severe impact on the individual transactions. You now want to turn from this negative thinking to more positive ideas. You want to define potential candidates for the long-term transactions. However, this is difficult to do. Most people have a difficult time being creative on the spot. Here is what we do. Use the following list items to serve as triggers for ideas an how to do the process. This should generate some interesting alternatives for the long-term transactions. A similar approach will be employed in the next chapter for Quick Wins and the future process through the improvement effort.

- Assume that money was no object. You can spend as much as you want.
- Take the organization out of the picture. Just assume that there are people working in the process.
- For the systems and technology, assume that they could support any way of doing the work.

If you consider the organization as being absent, then management factors, organization structure, and politics disappear. The point on systems and technology is important.

Next, compare the alternatives that you come up with and make a selection. You will want to wait until you have carried out the analysis for several transactions.

In our two example transactions, it is clear that more automation is needed for the first transaction, initial calls to the customers. In the ideal world, this transaction can be totally automated. In the second transaction, collection call, the systems can be modernized to online, intranet systems. The organization can be centralized into one center.

Define a Vision for the Long-Term Process

Now having developed models for several transactions, you can move up to the process level. Take the transactions that you developed and see what is in common. You can prepare a table where the rows are transactions and the columns are how the transactions are changed for the long-term process. Figure 6.6 presents a sample table. The areas of change need to be defined. Here is a list that can be used as a starting point.

- Systems
- Technology
- Policies
- Organization
- Procedures
- Staffing
- Facilities and infrastructure

Transactions	Systems	Technology	Policies	Organization	Processes	Staff	Facilities

Figure 6.6 Table of long-term transactions versus changes.

Compare the Current and Long-Term Process

You have now defined the long-term transactions and analyzed the current ones. It is useful now to make a comparison. It's time for another table. The example table in Figure 6.7 has four columns. The first lists the transactions, the second contains the key theme of the comparison, and the third contains comments. Our two examples are given here. It is possible to go from transactions up to the overall process, but this is best done by using descriptive text.

Estimate the Long-Term Future Process Scorecard

Here begin with the scorecard for the existing process developed in Chapter 5. Look at the tables and charts that were prepared in this chapter. Use these to define the new scorecard. You can use the scoring in Figure 6.4 as a guide. This will also serve as backup when you explain the scorecard for the long-term process.

Develop Improvement Tables and Process Scorecards

You can create improvement tables similar to those in Chapter 5 for the current processes. These can then be placed in the text of the process plan. Alternatively, they can serve as an appendix. To evaluate your work in creating the process plan, you can use the table in Figure 6.8.

Develop the Process Plan

The outline follows from the end product and milestone presented earlier in the chapter. Here are the key sections:

Current Process

- Brief description of the purpose and scope of the process
- Overall assessment of the process in terms of the process scorecard
- List of major issues and problems and their impacts, if not addressed

Transaction	Current vs. Long-term comparison	Comments
Initial call	Automation, policy	
Collection call	On line systems, centralization, organization	

Figure 6.7 Comparison of current and long-term future transactions.

Element of the Scorecard	Score	Comments
Involvement of employees in the		
process plan		
Extent of barriers addressed by the long-term process		
Realism of the long-term process		
Extent of differences between the current and long-term process		

Figure 6.8 Scorecard for the process plan work.

Long-Term Process

- Purpose and scope of the long-term process
- Description
- Comparison of example transactions
- Comparison tables involving the long-term process
- Benefits of the long-term process
- Factors that must be changed to attain the long-term process
- Potential scorecard of the long-term process
- Critical success factors to move toward the long-term process

Develop Enthusiasm for the Future

Although it may seem difficult to present a view of the future that gets other people interested, this step may be easier than you think if you center all reports and presentations on sample transactions. It is in the detail that people can see the benefits. It is very difficult to believe that you can go from here to there based on fuzzy text.

A word of warning: Experience reveals that people tend to be excited and want things to happen right away. You have to use the many constraints and barriers to dampen the enthusiasm.

What do you get from the enthusiasm? You can channel the support into helping resolve issues for implementing the Quick Wins later. This is specific and tangible and is, by itself, a sufficient reason for developing the process plans.

EXAMPLES

ASC Manufacturing

ASC management insisted that an overall plan for six major processes be developed. While this was useful, it turned out to be overkill. The work delayed the project and the plans turned out to be ponderous documents that few read. It was a case of too much, too late.

KOSAL BANK

The Kosal Bank example was presented throughout the chapter. In actual application, 12 transactions were employed in the development of the process plan. These represented over 90% of the work. The exceptions were noted, but they were not addressed. A critical success factor in later implementation was the support given by upper management as a result of the process plans.

HETSUN RETAILING

No process plans were developed. This had negative consequences. First, people questioned any change. Inconsistencies appeared and complaints about the lack of an overall plan surfaced. The team then developed an informal plan but never showed it to management.

LANSING COUNTY

The county management did not insist on any plans. However, the individual employees and supervisors wanted an overall approach presented. This was not a demand, but a request, so that they could understand what was going on. The plan was developed at the transaction level and presented. It served as a useful guide for the change.

LESSONS LEARNED

- Use this work to focus attention on policy problems and the need for reviewing specific business policies. Don't be vague. You should identify the policy and show its impact on the current work. Then you can describe the potential new policy and show how it benefits the long-term process.
- Organizational issues and problems are almost certain to be brought up and surface on their own. Acknowledge these concerns, but do not give them undue attention. People get more defensive about organizational issues than almost any other form of change.
- Try to develop the process plan in parallel with other activities so that it does not lengthen the schedule for process improvement. If substantial time is spent doing the process plan at the expense of other work, then the plan will begin to get too much scrutiny.

PROBLEMS YOU MIGHT ENCOUNTER

Potentially, the biggest problem that you might encounter is to raise expectations of employees and managers. They might begin to think that the long-term process will be implemented through Quick Wins and the initial process improvement. Obviously, this is not true. However, you should consider how to head off the potential problem. Here are some ideas:

- Make sure that everyone is aware of the constraints and barriers that prevent any immediate achievement of the long-term process. Emphasize both the business and IT aspects of the situation—not just IT.
- Use the tables that were developed in this chapter to show how different the current and long-term process versions are from each other.

WHAT TO DO NEXT

1. A good first step is to take a transaction in a process that you are familiar with and follow it through the actions that have been presented for analysis.
2. Following this, try to think of three or four creative alternatives for the long-term transaction.
3. Once you have completed these actions, then go back and add to the constraints.

CHAPTER 7

Define Your New Business Processes and Quick Wins

INTRODUCTION

You have the starting point defined—the current processes. You also have where you eventually want to go—the long-term processes. Now you must define a path for moving from one to the other. In developing the future transactions, you will be drawing on ideas of the employees, your industry data, and experience from using various software packages. In the end, you can change a process for the future and not the long-term by doing some combination of the following:

- Building, modifying, and buying software
- Implementing architecture and technology change on a limited scale
- Changing procedures and practices
- Modifying, creating, and eliminating policies

Now contrast this with Quick Wins. The scope of what is possible to change with Quick Wins must be limited to the following factors:

- Procedure and training material changes
- Minor policy changes
- Minor changes to the software
- Limited work on the architecture

There are some rules of thumb that we have followed in our improvement efforts. You should be able to implement Quick Wins in days or weeks. You should be able to implement the future process in months—not years. Experience has shown that if it takes longer than six months—or at the most one year—to implement the future process, you are likely to fail. Here is why:

- The process may deteriorate or change over this time.
- The internal business climate may change.
- Management may redirect resources away from the improvement effort to other work.
- New versions of the systems and technology that you selected may appear, making the technology and systems approach that you selected obsolete.

Any one of these setbacks can cause your plan to result in failure.

In Chapter 6 you defined the general shape of the long-term transactions. Then you extrapolated from these transactions to get a picture of the entire long-term process. This is not sufficient precision for process improvement plan that you are actually going to implement. You must consider dimensions of change for both Quick Wins and the future process. Consider the following dimensions:

- *Process change.* These are the changes to procedures and methods. This relates to "how."
- *Policy change.* These are the changes in the rules that govern the process. This relates to "how" and "what."
- *Organization.* There are any minor organization changes such as job descriptions that you will implement. This answers the question of "who?"
- *Infrastructure.* Infrastructure changes, like organizational change, have to be limited to the possible. This pertains to "what."
- *Resources.* This also relates to "who." Here you are identifying what resources are required to carry out the process improvement. This relates to both "who" and "what."
- *Other processes.* These are changes that you will make to surrounding and interfacing processes and work. This also pertains to "how."
- *Systems and technology.* These are the changes, additions, and deletions to support the new versions of the process. Systems and technology relate to "how."
- *Management.* This dimension refers to all aspects of planning, measurement, and control. This pertains to "why" and "what."

How do you employ these dimensions? First, you can employ them to help you generate ideas. Consider changing the process in specific dimensions. Second, you can employ the dimensions to understand how two alternatives for the new process are different from each other.

There are some caveats necessary here relating to the exceptions, workarounds, and shadow systems:

- Where possible, you should not address exceptions unless they are of sufficient volume, importance, and frequency. Doing so will take too much time.
- Workarounds are created when there is a problem with a policy or automation, typically. You will strive to resolve as many of these issues as possible to improve productivity.
- Shadow systems were established to address specific needs related to the work. You have to consider each shadow system to determine if (1) it can be eliminated, (2) it must be included in either the automation of the future or the procedures, (3) you can eliminate the need for the shadow system. Shadow systems get in the way of work overall often while helping the work in some of the detailed steps.

OBJECTIVES

TECHNICAL OBJECTIVES

The technical objective is to generate the new future process and the changes to the process for the Quick Wins. Your goal also includes having the actions taken in the Quick Wins lead up to the new process. You do not want to undo what you did and the benefits of the Quick Wins when you implement the new process. So they must be compatible with each other.

Another technical objective is to generate new transactions and processes that are feasible in terms of implementation. Here are some mistakes that people make:

- The transactions are too complex for the people doing the work.
- The manner in which different types of work is addressed is inconsistent.
- The transactions depend on such major changes to the systems and technology that it is impossible for them to be implemented in any reasonable manner.

Hopefully, you can head off these problems by remembering what you found in the technology and business analysis that you performed in the earlier chapters.

Another technical goal is that the new transactions in both Quick Wins and the new process must be sufficiently detailed to support implementation. If you remember the outline of the book, you can see that the next chapter begins the discussion of implementation planning. After defining the processes here, that work is completed.

For the Quick Wins you want to ensure consistency, but also you want to establish the basis for how the later future process is implemented. Putting in Quick Wins can build a pattern of successful implementation.

Business Objectives

While the technical objectives focused on reasonable completeness and validity, the business objective is to achieve tangible and not fuzzy or intangible benefits. You must convert any intangible benefits to tangible ones. Here are some examples to help:

- The new process is easy to perform. This means that training costs should be reduced. There should also be fewer errors or less rework. In addition, if the work is easier to perform, then people should be able to do more of the work in the same amount of time.
- The new process is simpler. This means fewer errors, more work for the same resources, and reduced training.

These are the normal business goals that you find in many reengineering and Six Sigma books. However, there is another goal. That is to have the changes and the new process be lasting and stable until you revisit the process to move it further along to the long-term process defined in the process plan presented in Chapter 6. This is so important that we included it in the title of the book. Thus, you have to achieve the following additional objectives:

- The process must be easy to measure. Hopefully, you can use the systems to produce productivity and other statistics. Ease of measurement means that you can detect problems and deviations from the new process quickly.
- The process and its transactions can be monitored and managed easily without extensive analysis and data collection.

Moving to Quick Wins the goal is to achieve tangible benefits in the short run that can help justify continuing to pour and apply resources to the improvement effort.

Political Objectives

The political objective for the future process is to achieve higher morale and involvement on the part of the employees. Through the Quick Wins you can establish confidence and trust in the improvement effort.

Another political goal is to handle resistance to change and any other political problems surrounding the process. You will likely not be able to change long-seated culture and political problems among departments, but you can get the process out of these problems to the extent possible.

END PRODUCTS

The general end products can be listed as follows:

- New transactions and process versions based on Quick Wins
- New transactions and the process for the future improved process
- Requirements and benefits to achieve Quick Wins and, separately, those to reach the future improved process
- Assessment of how far the future improved process took you toward the long-term process

There will be additional improvement tables as well. The requirements for the Quick Wins and future process define how to get from the current process to these. The benefits tell you what you will get out of the future process when you arrive.

Keep in mind that the new process definition must be very detailed. If the end products are vague, fuzzy, or have holes, then when implementation comes, people will start to invent fixes—potentially generating a worse process than the one you are replacing. This is exactly what happens with architecture. The architect designs a building. If the architect produces plans that are too general, then there will be no end of problems and issues.

There is also the other extreme. If you create an extremely detailed specification of how the process and its transactions will work that covers all situations, then there are other problems. First, you covered the exceptions. By doing this you gave them official status and now they are sanctioned. Second, this took too much time. The time would have been better spent in implementation.

METHODS

WHERE TO START

You begin by gathering the information that resulted from the previous steps of analyzing the current transactions, the business concerns, and the long-term process. We now return to the method that was discussed in Chapter 2 for organizing the work. As you recall, action teams were formed for each business area of potential interest. Each action team identified potential opportunities for change and improvement without regard to the time it would take to implement the change or each team's relationship with the other. You then applied this information in selecting which processes and transactions to go after in process improvement.

It is time to now carry the approach forward. You and the teams can sit down and assess the opportunities that pertain to the processes and transactions that were selected. Here are some steps to take:

- Review the writeup for each opportunity and have each action team vote on the opportunities for its own area.
- Having eliminated some opportunities, you can now receive input from all action teams on the remaining opportunities. This can be done through voting and through reviewing the documentation for each opportunity.
- The opportunities that pass these tests are then moved up to the steering committee.
- The steering committee provides input on the opportunities and selects the most promising for further analysis. These opportunities move back down to the action teams.
- The action teams now prepare more detailed business cases for each opportunity. The business case includes the problem that is being addressed, the impact if not implemented, the benefits, how the benefits will be verified, and what is required for implementation.

After you have carried this out, you have collected and refined the opportunities to get a complete list. In doing this, you also will likely define new opportunities that fill in gaps and holes. You will now begin to group the opportunities and define the new future processes and transactions.

Define New Transactions

Take the following actions to define the new transactions:

- Analyze and group the opportunities.
- Assess the impacts of each group of opportunities on the specific transactions. This will provide you with an idea of how the work would be performed if the opportunity were carried out.
- Define the transactions for the Quick Wins based on the opportunities.

Separately, but in parallel, you will work to define alternative future transactions. Then you can compare where the Quick Hits end and the alternative future transactions. You can then select the best alternative future transaction.

Let's consider each activity and provide some guidelines. How do you group opportunities? Here is a checklist for you to use:

- The opportunities complement each other and do not conflict. They can be implemented together. On the other hand, you can have opportuni-

ties where one requires the other to be in place. Then the trailing or following Quick Win is implemented in a later wave of Quick Wins.
- Two opportunities affect different but related processes in the process group in a positive way. They produce synergy so that they can be considered together.
- The opportunities can be carried out by the same resources so that they can be installed together.
- Both opportunities require some system, infrastructure, or policy change that precedes them. The change goes in the first round of Quick Wins and the opportunities can go in the next wave.

Uncover Quick Wins

Next, you will seek to determine the impact of implementing the group of opportunities on the process. You will want to answer the following questions:

- Can the employees in the department handle these changes at one time? The same question can be applied to customers or suppliers in e-Business.
- Do you have the resources to effectively carry out the group of changes?
- What disruption will occur in the business process during the transition and implementation?
- Will it be possible to measure the results of implementing of the group of changes?
- What happens if there are problems? What is the contingency plan? Do you just fall back to the original process?

With the impact understood, you can use the grouping to generate the Quick Wins that are combinations of opportunities that passed the preceding tests and analysis, which ensured that they were complete.

How do you determine whether the Quick Wins make sense and assess their ease of implementation? Here are some guidelines. First, consider the following list to assess what is necessary for implementation:

- Minor system changes
- Infrastructure, office layout, and facilities changes
- Redoing and changing procedure and operations manuals
- Setting up any files, forms, logs, or other items to support the Quick Wins
- Training the employees in the new Quick Win version of the process
- Monitoring the initial operation of the Quick Win transactions

You can also build a transaction table. Use the example given in Figure 7.1 to list the steps and add two new columns. The first is the difference between the

Transaction: ______________________

Step	Who	What	Difference from current process	Issues and comments

Figure 7.1 Quick Win version of a transaction.

Quick Win version and the current transaction at the step level. The second highlights any issues and allows for comments. You will do the same thing for later versions of Quick Wins by comparing them to the previous version.

As you can see from the work you did to complete the table, the extent to which this detailed analysis can be done is limited by time and resources. Developing the table in Figure 7.1 has the side benefit of leading to the new operations procedures and training materials, however, so you can argue that it would have to be done in any event. Another observation relates to the exceptions that have been discussed several times. You can see that if you had to develop Quick Wins for the exceptions, you would become buried and the improvement effort would likely end here.

Determine the Future Process

While you are doing the work on the Quick Wins, you can begin to consider generating the transactions for the future process. Experience shows that creativity is difficult to exercise under pressure. Therefore, the guideline is to consider various triggers and other factors that get you and the team thinking about new ways to do the work. Here is a list of triggers that have been useful in generating ideas for the new process.

- *Eliminate the transaction.* This is obviously radical, but should be considered. How would you replace it? This trigger fits in the management dimension. It is an extreme step, but it is worth considering.
- *Place the transaction outside of the organization.* This includes outsourcing that fits into both the management and organization dimensions. This is valid for transactions that are generic and involve no extensive organization-specific knowledge.
- *Starve the transaction.* Deny the resources to maintain and enhance the work. Let the transaction die. Eventually, the transaction can be eliminated. This fits with the resource and management dimensions. Politically, we have used this as a step toward elimination.
- *Merge the transaction with other transactions.* This is quite common. You could

divide the transaction and apportion it among other transactions. This fits in the other processes dimension.

- *Transfer the transaction to another organization.* The transfer would be done so that the transaction is intact. This relates to the resource and organization dimensions.
- *Throw people at the process.* This sounds absurd. However, it is worth considering what you get when you apply more resources. You will, for example, overcome any problems with resource shortages.
- *Start with management measures and work backward.* Here you begin with performance measures and statistics for the process and transactions (what you are trying to achieve). This is a variation of the "work backward from outputs" approach. Here you are working from the top down. At the heart of this trigger, you are considering the importance of the transaction.
- *Throw money at the work.* What can money buy? Well, think about it. Throwing money at a process or work often makes the process less efficient and effective. Money can bring various resources into the work. This trigger helps you to consider priorities among processes for resources. The analysis here can also reveal problems that cannot be solved with money and resources—policy problems, organizational issues, and interfaces with customers and suppliers.
- *Move data capture up in the process.* This applies to transactions where data are being captured. Moving the capture up usually reduces paper and manual effort. It can also reduce errors.
- *Copy the best competitor.* Use the information gathered in Chapter 3. This trigger can reveal the goodness-of-fit, but what the competitor does never seems to fit exactly. Attempt to discover why and you will find some of the factors that differentiate you from the competition.
- *Outsource the work.* This is more specific than the earlier trigger. It is a good test to see if the work is generic or specific to the company. If it is specific, then you can ask, why? The answer may reveal other things to fix in the process.
- *Downsize the resources.* You can reduce resources being fed into the process. You can also downsize the systems from mainframe computers to PC networks. If downsizing appears to work, then the problems with the current process may not be structural.
- *Abolish the organization.* Here you turn your attention to the organization and away from the process. What if you could do away with the organization and start fresh? What would happen to the process? Who would do the work? This type of radical trigger also helps to define the boundaries of the process and transaction.
- *Split up the transaction.* Can you split up the transaction? This tests your knowledge of the surrounding processes and work.
- *Eliminate paper and forms.* Eliminating paper and forms supports the analysis of simplification, automation, and work elimination.
- *Automate all of the transaction.* Here you determine how much of the

transaction can be automated. This analysis reveals what has to remain manual. An example might be where there is some judgment needed. Some examples where automation is limited are customer service and supplier contacts.

- *Eliminate all automation and replace it with manual steps.* This is the opposite of the preceding trigger and indicates the severity of the automation issues on the process. Manual processes are also very flexible.
- *Perform the transaction with different people.* Maybe the issue is not the organization, but the specific individuals doing the work. If you replace them, how long would it take to bring new people up to speed? Are the documentation and procedures sufficient to support the new staff? What part do personality and personal relationships play in the performance or lack of performance of the work? This trigger is important because many processes depend on key employees who have been doing the same work for many years. A small core of individuals may be all that is keeping the work going.
- *Move the transaction to suppliers and customers.* From automated teller machines to telephone and Internet self-service, this has been the option of choice for many situations. Not all processes are suited to this trigger. It would be difficult to move all of accounts payable out to vendors, for example.
- *Change the policies governing the transactions.* This trigger should be considered more often. Policies are great opportunities since they can be changed or eliminated at the stroke of a pen. You can see clearly the impact of the policy by using this trigger.
- *Break down the transaction.* Here you can simplify the transaction and break it down into parts. The work is reorganized into smaller and simpler parts. A less-specialized staff can handle the mundane parts of the transaction.
- *Open up information to all in the group.* In some processes, people keep information to themselves. People do this to retain power. When you open up the information to all, you depoliticize the work. Then you can determine the impact of politics and hidden agendas.
- *Have fewer people do the work.* This trigger is useful to enforce accountability and control. If you consider having one person do the entire transaction, then you can determine where the work is complex and where specialized knowledge is required.
- *Change the location where the work is performed.* This is an infrastructure trigger. Location can play a role in process performance. Here you can consider centralizing the work.
- *Minimize or maximize customer contact.* Here you could evaluate the effects of having more information in the customer contact. This might improve performance of the work.

Of course, these triggers are being applied at the transaction level. For some small processes, you might want to apply the same triggers for the entire process.

You can now create a table in which the rows are the applicable triggers and the columns are the dimensions listed in the introduction of this chapter. In the table presented in Figure 7.2, you can place an X in an entry if the trigger requires the specific dimension.

Other approaches are useful for generating ideas for the new transactions. Here are five of these:

- *Bottom up*. Start with a specific transaction and improve each step in the transaction. Then move up to the transaction and then to the process. This will likely yield valid changes, but may not result in major change.
- *Top down*. Start with the process and divide it into parts or subprocesses. Move down to the transactions. This may give more flexibility with the process and preserve the boundaries of the process.
- *Infrastructure and architecture*. Address the processes from the standpoint of fixing the architecture and systems issues. Then see what impact this has. This will reveal the limitations of the technology.
- *Outside in*. Begin with the customer or supplier. This is how Six Sigma is commonly carried out. This approach gives you a clear idea of the customer interfaces. However, it does not cover much of the internal part of the processes.
- *Organization based*. Here you would work on organizational issues and determine where the process should be placed in the organization.

Using the triggers and the preceding approaches, you can begin to create alternative ways of doing the transaction.

You can now organize how the group of processes would be performed. You can start to evaluate the alternatives for the future processes. Here are some actions to take:

- Eliminate some alternative by comparing them to the current methods of doing the work. You can determine if any alternative is only marginally better than what exists. Or, you might see that an alternative violates the constraints defined in Chapter 6. A technology or infrastructure investment might be too large or require too much time to be feasible.
- Aggregate similar alternatives into sets. Consider the attributes of the collection or set of alternatives.

Dimensions

Applicable Triggers	Process change	Policy change	Organization	Infrastructure	Resources	Other processes	Systems & Technology	Management

Figure 7.2 Triggers for the new process versus dimensions.

In general, you will proceed by elimination. This is reasonable since you will not be developing more than three to five alternatives. Most of the time you will easily get down to one or two alternatives. You can now take the finalists and create more improvement tables:

- *Future transactions versus the improved architecture.* This is the fit of the future process to the new architecture that will be implemented. A lack of fit indicates limited benefits of the technology and that the new process is not taking advantage of the new technology and systems.
- *Future transactions versus the organization.* This table indicates the degree of fit between the future process and the current organization. Lack of fit indicates that there may have to be substantial organizational change.
- *Future transactions versus the process issues.* Here you can see if the future process addresses the known issues with the current process.

You don't have to stop here. You can construct the corresponding tables that you did for the current and long-term transactions. These are listed in Appendix 2.

Now move up to the future process group level and you can generate more improvement tables:

- *Future process group versus business objectives.* The table indicates the degree to which the future processes meet the business objectives.
- *Future process group versus business issues.* The entry is the extent to which the future processes address business issues.
- *Future process group versus the new IT architecture.* Here you can see the degree to which the processes take advantage of the new systems and technology.

As a final step, consider simulating how the new processes will work with the action teams. You can do this manually at the level of individual transactions. You first begin with reviewing the current transaction. Also examine the issues and problems in the process. Now walk through the future transaction. Evaluate it in terms of benefits and whether the issues were addressed. This step is very useful, since it will likely surface implementation issues.

Identify the Process Road Map

This step is to create an overall view of how you are going to move from the current process to the new process via the Quick Win versions of the transactions. The purpose of this action is to show the action teams and management how the overall picture looks and to seek additional input.

At this point you have created the Quick Wins, the new future process, the analysis of these, and the process road map. You also have the business cases for the opportunities that support the Quick Wins. You are ready to carry out the following steps:

- Action teams vote on the business cases and transactions and processes.
- The results then move up to the steering committee.
- After voting by the steering committee, you proceed to the executive committee for review and approval.

What are the management committees approving? They are giving the go-ahead to begin implementation planning for both the Quick Wins and future processes. Your next steps will be to develop the detailed implementation strategy, organize the implementation effort, and develop the implementation plan (Chapters 8 through 10, respectively).

DEVELOP THE FUTURE PROCESS SCORECARD

The improvement tables have already been covered. Let's turn to the evaluation of how you are doing in defining the future processes. The table that you want to prepare is given in Figure 7.3.

DEFINE QUICK WIN SCORECARDS

Here you can employ the table in Figure 7.4. Note that a number of the factors are common with that of the future process in Figure 7.3.

Element of the Scorecard	Score	Comments
Level of participation of action teams		
Consistency with the long-term processes		
Involvement of management and employees		
Extent and range of alternatives considered		
Disruption and risk to the process		

Figure 7.3 Scorecard for future process definition.

Element of the Scorecard	Score	Comments
Level of participation of action teams		
Consistency among the Quick Wins		
Extent of potential benefits of Quick Wins		
Consistency with the future processes		
Involvement of management and employees		
Opportunities dropped and their benefits		
Disruption and risk to the process		

Figure 7.4 Scorecard for the development of Quick Wins.

EXAMPLES

ASC Manufacturing

Quick Wins were thought initially to be few and far between. Yet when department meetings were held, literally hundreds of suggestions were made. Most of these were useful and could be implemented independently of the new processes and systems. These stand-alone improvements are valuable since you can establish momentum for change and support without interfering with the implementation of the future processes and systems.

Kosal Bank

Here were some of the Quick Wins that were identified:

- Change of office layout
- Change in organizational structure for collectors so that there were separate groups for tracking people down (called "skip tracing"), collections, and handling difficult and persistent collection customers
- Consolidation of the collection activity into 11 centers
- Implementation of formal training and procedures for each job type

For the future process, the following were key ingredients:

- One queue of delinquent accounts so that the most delinquent account was handled first.
- Automated management statistics to be able to address both productivity and effectiveness.

- Elimination of all paper records and forms. All information was stored online.

After implementation of the system across the 11 centers, there was sufficient standardization and control to support the further consolidation into one center—resulting in further savings and benefits.

Chapter 5 identified several issues related to the collection effort, including recording manual results and reviewing logs. In Chapter 6 the long-term transaction was almost entirely automated. The Quick Wins move the transaction to becoming more standardized and get it ready for automation. The future transaction then provides for limited online operations.

HETSUN RETAILING

More than 200 Quick Wins were identified. Space does not permit even a portion of the list to be printed here. The Quick Hits were spread across accounting, finance, marketing, sales, IT, inventory control, vendor relations, and other areas. The future process was based on a single intranet, Web-based system for internal employees, retail customers on the Web, and international sales and customer service. This then required substantial modifications to procedures and policies so that the system could be developed and implemented in a limited time.

LANSING COUNTY

In the operations area, more than 50 Quick Wins were defined. These ranged from job simplification and the elimination of forms and paper to changed policies and procedures and the development of temporary shadow systems in the department. These were and did serve as prototypes for the future systems. However, they were working systems based on the database management system Microsoft Access.

LESSONS LEARNED

- Suppose that you are evaluating several software packages to implement so that you can improve a process. Package A has many more features than package B. Both are priced similarly and use similar hardware and system software. However, package B has greater functionality in that without being changed it will handle 20% more transactions than the other package will. Which package do you choose? The traditional view is to select package A since it has more features. Features do not mean as much as functions, however. Features are nice to have. Functions are must haves. You should pick package B.

- Remember to initiate the work to define the future process while you are doing the work in Chapters 5 and 6. You don't have time to treat this work sequentially.
- Do not assume that you can introduce the Quick Wins informally. You can, if the group is small and the changes are minor. Otherwise, you will want to be more formal and develop more documentation. The consequences of not doing this are that the old process will likely continue and the Quick Wins will just overlay the old process—causing more work and lost productivity.

PROBLEMS YOU MIGHT ENCOUNTER

- A common mistake is to jump too soon for the first version of a new process that appears. The most frequent form of this mistake is to build processes around specific software systems. The concept is that if you warp the transactions around and through some software—that will be the best solution. The rationale is that, after all, the software works at many other places. Software vendors often indicate that this will work. However, the problem with software packages is that it is not what you get with the software that is important, it is what you did not get. You will have to invent new shadow systems and do customization to make up for missing functions.
- It can also happen that you are unable to define good candidates for the future processes due to an excessive number of constraints and restrictions. This can be a plus since you can use this drawback to show management the impact of the constraints on potential changes.

WHAT TO DO NEXT

1. Take the transaction that you analyzed in Chapter 5 and then worked with in Chapter 6 for the long-term process. Follow the actions in this chapter to define at least three alternative new transactions.
2. Carry out the analysis of the three alternatives and assess which best works to achieve the company's technical, business, and political goals.
3. Now back off from the new transaction and try to define one or two stages of Quick Wins that will help to implement the new transaction.

PART III

Plan for Process Implementation

CHAPTER 8

Develop Your Improvement Implementation Strategy

INTRODUCTION

At this point you have defined the basis for implementing process improvement. More specifically, you have performed the following work:

- Selected the specific processes and transactions to address
- Determined the problems, issues, and opportunities for the current processes
- Defined the long-term processes to establish targets
- Developed the future processes that take you from the current situation toward the long-term processes
- Identified the best Quick Wins to help you achieve the future processes

So now you just implement the process, right? Wrong. If you plunge in and start making changes, there could be more problems than the ones you are attempting to address. This is the problem with some implementations of Six Sigma and the problem with reengineering. Management often sees what is to be done and tries to have it done tactically. However, process improvement is neither routine nor simple. It is complex due to various political, cultural, business, and technical factors.

In this chapter you will develop a road map that sets out the sequencing of the Quick Wins and the future processes. In Chapter 9 you will define a project plan to support this road map. The road map or improvement implementation strategy lays out how to go from the current situation to the future processes. It gives the big picture.

What are some of the problems that will likely occur if you avoid defining the road map? Here are some we have observed:

- Management will want to speed up the rate of change. You know that this does not make sense, but you have nothing to demonstrate to management that this is not a good idea. The road map can clearly demonstrate the impact as the example reveals.
- There may be resistance to some of the Quick Wins. To overcome and address this, you can show people how the changes go together.
- People can become confused or depressed. They have seen all of the analysis results and wonder what will happen. Without a road map you have to explain repeatedly what will happen.
- The improvement team will develop a detailed project plan for process improvement. This will not be clear to anyone but members of the team. No one can make the leap from the concepts and ideas of improvement to the plan.
- There is no opportunity to do trade-offs on sequencing. Instead, management is left with a detailed implementation plan that can only be addressed through completion dates, rather what is to be done.
- You need to think about how the Quick Wins will group. Therefore, you must have a vehicle to communicate with people. The road map does this.
- Without the road map you have no easy way to chart progress and results. All that you can do is to issue documents that few may read.

Why don't people develop the strategy? Here are some common reasons.

- There is the fear of doing extra work. The idea is that if you spend the time doing the road map, then the work will be slowed down. In fact, the opposite has proven to be true. You can shorten the time to develop the plan with the road map. Next, you will have to organize the Quick Wins into groups later. Without analysis and trade-offs, you will make mistakes. These will turn out to be costly in both time and effort.
- Another fear is that the true scope of the improvement effort will be revealed. This will then raise the level of resistance to change. This is true if you lack support for the improvement effort. However, it will then just delay the inevitable—failure.
- People don't think about this. They treat process improvement as a standard IT project. They want defined requirements. These were developed in Chapters 5 to 7. However, there are requirements and there are requirements. The ones developed in the earlier chapters were tactical requirements. The requirements that you are building here are strategic requirements. You are indicating the overall sequencing of change.

Now turn to the flip side. What are the benefits of having the strategy?

- The strategy provides a link between the business mission, vision, objectives, and the improvement effort.
- You can develop alternative groupings of Quick Wins in different road maps and then assess their impacts.
- By seeing the big picture, you can head off management changing direction and setting new priorities that disrupt the work.
- You can use the improvement implementation strategy to obtain resources for the improvement effort.
- You can perform trade-offs in terms of speeding things up or slowing them down.
- It is useful as a political weapon in dealing with resistance to change.
- With the road map, you can decide that the technology or marketplace is not right for the implementation of the future process. Then you can stop for a while with the Quick Wins. At least you have something to show for the effort. With the standard system implementation approach, it is all or nothing. Too many situations end with nothing.

Let's define the improvement implementation strategy in more detail. At the heart is a table. As you have seen, we have frequently resorted to tables, lists, charts, and diagrams to document, analyze, and market process improvement. Massive amounts of text in general and a detailed strategy report in particular don't make much of an impact. No one has the time or interest to plow some arcane document. The general framework of the table is shown in Figure 8.1. Let's consider this diagram carefully:

- The first column contains the areas of change (e.g., IT, procedures, organization, policies, marketing, customer service) as well as areas of impact.
- Impacts include costs, benefits to the firm, risks, and the benefits to the target audience. In essence, the impacts sum up what is in the area entries of the table.
- The current situation describes what is going on now. In these areas you would place issues. Impacts would be the effects of the problems and issues.
- In all but the last column, you have each phase of Quick Wins. Figure 8.1 shows two phases of Quick Wins—typical of many situations.
- In the last column you always put the future process.
- The area entries in the Quick Win columns are the changes that you will introduce in each phase.
- The area entries in the future process column describe and summarize what is going to be done to implement the future process.
- The impacts for the Quick Wins and the future process columns are summaries of what happens when you implement the changes.

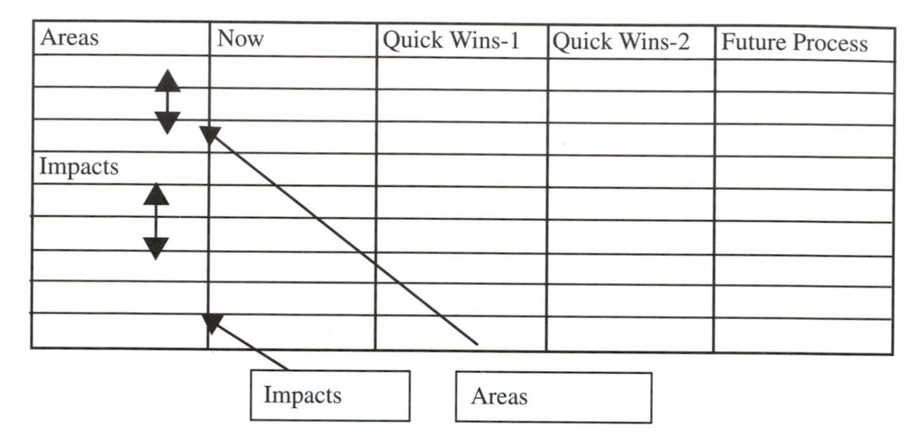

Figure 8.1 General framework of the improvement implementation strategy.

You will want to develop a table for some specific transactions. This will get you into the method. Then you can generalize to a specific process. After doing this for each of the major factors in the group, you will then be able to develop a summary improvement implementation strategy table.

Let's take a simple example. Suppose that your organization has a simple web page. It doesn't get many visitors. Management has decided that they want to get more involved in e-business. They want to replace the web site with something that does real work.

Let's first define the objectives and scope. Rather than just another improved web page, the technical goal is to implement transactions that will support customers doing more functions. The business goal is to improve the internal processes that relate to the customer as well as to support the new web page. The political goal is to get employees and managers in the organization working together. The scope is then process improvement with the target of serving customers better.

In doing the analysis, you find that there are many problems with the internal processes—resembling those in the examples in Chapter 5. The long-term processes involve the customer being able to do most transactions on the Web. The future processes aim at providing for customer ordering, follow-up on orders, payment, and canceling orders (Chapter 7). Figure 8.2 contains a potential overall implementation strategy. Note that this table is an oversimplification. In real life you would add more areas, such as customer service.

Keep the following considerations in mind when completing the table:

- You will be implementing two phases of Quick Wins. The first will standardize the procedures; the second will employ lessons learned to better sell and service the products.

Areas	Now	Quick Wins-1	Quick Wins-2	Future Process
IT	Client-server ordering	Prepare network; system development	System development and testing	System implementation
Order processing	Customer service center	Standardize procedures; eliminate exceptions	Gather lessons learned	Implement improved business rules
Marketing	Batch oriented	Review how others sell on the Web	Design on line marketing and sales campaigns	Implement on line sales campaigns
Impacts				
Costs	Labor intensive	Analysis, training, IT	IT, labor costs	Higher support costs
Benefits	Same methods for years	Increased productivity	Improved product information	Increased sales, productivity
Risks	Inefficient; not competitive	Lack of impact	Lack of real improvement	Lack of customer use
What customers get	Limited method of purchasing			

Figure 8.2 Implementation strategy for an improved Web site.

- The IT approach is to do internal system development. The goal is to implement one standard Internet, Web-based system for employees and customers.
- The benefits are improved productivity and then reduced costs as customers do more via the Internet. The costs are the normal costs of IT and labor.
- The risks are that there are minimal impacts until the future process is installed. Then the risk is that consumers will not use the site.

There are some interesting side notes to this example. First, it demonstrates to management that process improvement and, in this case, e-business are mostly business tasks and work. However, in most cases, the costs of the improvement effort lie in IT. This is an example of the basic dilemma of a business project with IT costs. That is probably why so many e-business efforts are viewed as IT projects.

The second thing to get from this example is that there are no times or schedules given. All that you have are phases of work. There are no dates. This is important in understanding the improvement implementation strategy. The road map tells you what will be done. It does not tell you when it will be done. That is the focus of Chapter 9, where the schedule is determined.

A third important observation is that each entry in the areas of change in the upper part of the road map is a project. This is very valuable because it reveals how you might best organize the work.

In the example you can also see how you can perform trade-offs:

- Let's suppose that you want to speed up the IT work. That sounds good. However, it is unlikely that either the lessons learned or marketing will be ready to do much. The benefit here is that you will develop a more realistic schedule that will sell to management better.
- Now assume that some area is going to slip its schedule. This means that the column entries of that row shift to the right. If this occurs, you might consider slipping the entire schedule since pieces of the future process will be in place for an extended time without others.

OBJECTIVES

Technical Objectives

The major technical goals are to group the Quick Wins in a logical and feasible way and to then indicate how these lead up to the future process. Other technical goals are the following:

- Develop the basis for the project plan.
- Understand where the risk is and the issues are so that they can be addressed as you proceed.
- Define in a complete manner the changes that are required to lead up to the future process.

Business Objectives

A business objective is to ensure that management and employees have an understanding of what is going to happen. By being involved in the development of the table, the business can support the improvement effort better.

A second and more significant goal is to ensure that there are business benefits to the Quick Wins. This requires that there be sufficient attention to the changes that will be introduced.

Political Objectives

While many people may have supported the improvement effort so far, a certain number probably felt that nothing would happen. After all, there have been other projects to make things better in the past and nothing came of them. With the improvement implementation strategy you are showing that this is getting serious.

The development of the road map also will serve to surface other points of resistance to change. Again, it is like system testing—the more errors and the sooner they appear, the better.

We have stated all through the book that a goal of improvement is to make people's working lives better through the changes. The table is an excellent way to get people involved and to show them how they will benefit.

END PRODUCTS

The basic end products are a series of improvement implementation strategy tables at the transaction, process, and process group levels. These will serve as the basis for implementation.

Additional end products can be developed. First, you can use the road maps to demonstrate how specific issues in the current processes are going to be addressed. Second, you can demonstrate how the phases of Quick Wins and the future process support the business mission, vision, and objectives of the organization.

METHODS

Where to Start

The overall method is as follows:

- Define the structure of the improvement implementation strategy tables.
- Create alternative tables and road maps for managers and employees to discuss and review. This will help sharpen people's understanding of the changes ahead.
- Evaluate and select the winning road maps for each process and the process group.
- Market the final road maps to management and employees.

Let's get going. You will want to have the Quick Hits and the future process close at hand.

Determine Areas of Change

In order to build the table, you will first want to identify the areas that will change. You will be grouping the Quick Wins and the elements of the requirements for the future process. Experience has shown that several groupings are possible:

- Group overall so that multiple business areas are in one row. This is probably most useful at the level of the process group. It is too general for lower levels.
- Group by function. That is, identify the business areas, such as finance, marketing, sales, accounting, IT, customer service, and vendor relations, and use these as rows.

Try to limit the areas to five or fewer for any one phase. If you have to create a large number of rows, then the scope of the Quick Wins is probably too large and must be cut down to ensure success.

For the impact areas, you might start with the ones in the example discussed earlier in this chapter: cost, benefits, risk, and the benefits to the audience. You can select employees, customers, or suppliers. What about management benefits? These are in the benefits row.

Define the Current Situation for the Road Map

This is the second column. For each area you will return to your work in Chapter 5 to identify some of the key issues. Try to capture the essence of the problems in one or two bullets. You might want to make these fairly provocative so that you will get some reaction. Politically, it will be an advantage to have people once again understand the problems that exist.

Now you proceed to the lower part of the column in the areas of impact. Rather than ask people what the impacts are, use the improvement team and the action teams to get their input.

It is now time to reveal the table as it exists so far to management and to selected employees. Here you are attempting to achieve several goals. First, you are validating the current situation and its impacts. Second, you are giving people the opportunity to understand the table and its structure. Experience shows that if you wait until you have the alternative road maps, there will be difficulties because people are trying to understand the structure and the content at the same time—not a good idea.

Construct Potential Improvement Implementation Strategies

After getting feedback and suggestions, you are now going to group and organize the Quick Wins by phases. This can be a daunting effort if you don't follow an organized approach. Here are suggestions.

- Proceed by rows. That is, you will first identify the Quick Wins for each area. Leave the future process for later.
- For each row, sit down with the appropriate improvement team and action team members to identify dependencies between Quick Wins. You will find that some should be done together at the same time so as not disrupt the departments excessively.
- Try out different alternatives for each row, including the following:
  - — Implementing all Quick Wins implemented as soon as possible.
  - — Stretching out Quick Wins to as late as possible.
  - — Implementing Quick Wins for specific transactions in groups.
  - — Installing Quick Wins that involve management policy last or first.
- You have now defined several alternatives for each row. Now try to create groups of rows based on the same philosophy as the previous action. You are likely to run into limitations due to resources that are available.
- The result up to this point is series of partially filled tables. The next action is to determine the impacts of each phase of Quick Wins.

Assuming that you have done this for several transactions and for each major process in the process group, you can now get feedback from the action teams. There are likely to be many arguments about the benefits and risks. For these you can return to the business cases you created for the Quick Wins in Chapter 7.

It is now time to consider the future processes. You will first define what is needed in each area—doing this as was done for the Quick Wins. Now stop! Before you can define the impacts, you will need to determine the stretch or distance between the last phase of Quick Wins and the future process. Doing this will help in the evaluation of the alternatives. If there is too big a leap, then there are insufficient Quick Wins. Alternatively, the future process changes may be too ambitious. You desire to select a strategy that flows as smoothly as possible from the Quick Wins.

The next action is to determine the impact entries in the table for the future process. Here you can return to your work in Chapter 7 for help.

It is time for review and feedback by the action teams and improvement team. There is likely to be an interesting discussion on the impacts as well as renewed discussion on the Quick Wins. After all, people are seeing all of the analysis summarized in one place at one time.

EVALUATE AND SELECT YOUR IMPROVEMENT IMPLEMENTATION STRATEGY

A number of alternative strategies have been developed at the transaction, process, and process group level. It is time for evaluation and selection. We suggest that you proceed by elimination. Here are some criteria for elimination:

- The impacts of the alternative indicate that too much of the cost is up front and the benefits are later. Not only will management not go for this, but also the employees will wonder what is going on.
- The changes between processes are not consistent in each phase. You are doing a lot for one process and next to nothing in another. This will create an uneven workload for the employees and action teams.
- There is too large a leap from the Quick Wins to the future process for an alternative. This will make the future process more complex, difficult, and risky to implement according to a schedule.
- An alternative uses too many resources in some areas and too few in others. There should be more resource leveling.

You want to go into the analysis with the action teams. Consider a voting system in which the team members employ a scorecard method. Figure 8.3 gives an example.

Following the selection of the strategy and review by action teams, you can proceed to the steering committee. For this committee you will want to have an organized presentation. Here is one that has worked repeatedly:

- Introduce the improvement implementation strategy and what it is intended to achieve. This provides the basis for what is to come.
- Present again the current situation. This reinforces the problems and awareness of them.
- Show the overall table without the impact area rows. Go over the Quick Wins and future process in each row. Then you can discuss the consistency of the column. Here you can also outline the evaluation method for arriving at the strategy. You might even present a losing strategy as an example.
- With the understanding of the changes, you can now move to the impact rows. It is suggested that you present each column.

Prior to the presentation it is a good idea to present and walk through the table with selected managers in each area. You will want to explain their row in detail after you have shown them the big picture.

There is likely to be a great deal of discussion on the schedule and timing of the work. Take down any suggestions and comments. However, indicate that the main goal is to determine the "what," "how," "who," "how much," and "where."

Alternative: ____________________

Scoring Element	Score	Comments
Feasibility in terms of resources		
Likelihood of resistance		
Reasonable achievement of benefits		
Allocation of costs across phases and future process		
Burden of effort on department employees		
Potential risks to customers, employees, or suppliers		
Ability to manage the change in each phase of Quick Wins		
Potential problems that might arise		
Consistency of the Quick Wins in each phase		

Figure 8.3 Example of a scoring table for alternative improvement implementation strategies.

The "when" will follow soon. You will follow the same approach for the executive committee. Be prepared to indicate when you will be coming back with the implementation plan.

Construct Improvement Tables

In previous chapters you developed a series of improvement tables. These are summarized in Appendix 2. At this point it is good if you prepare some additional tables to be able to link the improvement implementation strategy to the previous work. Here are some good ones to use. Note that there are too many to use them all. Start with a few and then review these. You may want to incorporate them into the presentation of the improvement implementation strategy to management. The areas of the Quick Wins and future process are the rows in each table. The entries can either be a numerical score (e.g., 1 to 5) or comments:

- *Quick Wins in a phase versus business mission.* This table shows how the Quick Wins provide support for each element of the mission.
- *Quick Wins in a phase versus business vision.* The vision is the long-term view of the business. This table reveals how the Quick Wins by area contribute to the vision.

- *Quick Wins in a phase versus business objectives.* The support of the objectives by the Quick Wins are shown here.
- *Quick Wins in a phase versus business issues.* The degree to which the Quick Wins contribute to the resolution of the business issues is shown in this table.
- *Quick Wins in a phase versus current processes.* This table shows the impact of a group of Quick Wins on each current process in the process group.
- *Quick Wins in a phase versus future processes.* The entry is the degree to which achieving the Quick Win gets you closer to the future process in the group.
- *Quick Wins in a phase versus long-term processes.* The table reveals the extent to which Quick Wins in each area contribute to achieving the long-term process.
- *Quick Wins in a phase versus process issues.* The entry is the contribution of the Quick Wins to resolving the specific issues in the business process.

You can generate similar tables for the future process. These are listed here:

- Future process changes versus business mission
- Future process changes versus business vision
- Future process changes versus business objectives
- Future process changes versus business issues
- Future process changes versus current processes
- Future process changes versus long-term processes
- Future process changes versus process issues

Develop the Implementation Strategy Scorecard

Figure 8.4 gives a suggested scorecard for assessing your effort for the improvement implementation strategy. You might want to have the improvement and action teams participate in this evaluation. This is a good way to have people start doing measurements prior to implementation.

Assemble the Implementation Team

It is not necessarily true that the same people who participated in the analysis so far should be involved in implementation. This also applies to the improvement leaders. That is why we said at the beginning that it is valuable to have two project leaders. Up until now, one of them served as the overall leader. It is now time to consider a hand-off to someone who is more implementation focused. This is not meant to be negative. People are different. Some are good at analysis; others are good at getting things done. You want a manager who will take the lead in developing the improvement implementation plan in the next chapter.

Element of the Scorecard	Score	Comments
Degree of involvement by action teams		
Number and range of alternative strategies generated		
Validation of costs		
Validation of benefits		
Validation of risks		
Validation of benefits to customers, etc.		
Reasonable of workload level in each phase		
Elapsed time to prepare the strategy		
Extent of work done on the implementation plan		

Figure 8.4 Scorecard for improvement implementation strategy.

Now turn to the team. Review the Quick Wins and future process actions. You can determine in which areas you will need involvement and support for implementation. Here are some guidelines for selecting team members:

- Keep in mind that implementation of improvement usually requires limited periods of full-time effort. Being on the action teams was a part-time responsibility. Between spurts of involvement the employees can return to their normal work.
- Try to avoid individuals who are key to doing work in the department. They are handling exceptions and helping others. If you pull them out, the impact on the department can be disastrous. You can get their input and have them involved in training, but not on a full-time basis.
- Look for who is available to work on the implementation. Selecting someone who has many other priorities will cause more work and raise political issues as you fight for that individual's time.
- Consult with action team members to identify suitable candidates.
- Consider individuals who are junior and who supported change in the work done in Chapter 5.

How to employ the team members will be discussed in the next chapter.

Define Culture and Change Goals

Beyond the technical and business goals of the improvement implementation strategy, you should consider what is to be accomplished from a political and cultural standpoint. Here you should review the road map and determine what changes will occur in these areas during implementation. A major goal we think is to instill more sharing of information and collaboration.

Develop the Implementation Organization Scorecard

Here is a chance for another scorecard. Figure 8.5 can be employed as a starting point. Here you are testing yourself in terms of how you went about selecting team members for the improvement implementation.

EXAMPLES

The tables for the example firms are too large to discuss here. For our standard examples, we discuss the experience of developing the road map. A new section of examples has been added to give you some more specific examples.

ASC Manufacturing

At ASC, there was originally no effort to build the improvement implementation strategy. Instead, the detailed project plan was developed. It involved more than 1,500 tasks. This drove people nuts, as no one wanted to take the time to understand it. The team backed off of the plan and went back to develop the strategy. This was successful. Then the team thought that it could use the previously developed plan. It turned out that the old plan was inconsistent with the new strategy. A new implementation plan had to be developed.

Element of the Scorecard	Score	Comments
Number of people considered for the team		
Size of the team		
Project leader selected		
Elapsed time to identify the team		
Identified conflicts with other work among team members		
Number of people who turned down being on the team		

Figure 8.5 Scorecard for improvement implementation team.

Kosal Bank

At Kosal Bank, an overall implementation strategy was created. This was different from the ones in the chapter as it did not address benefits, costs, and other factors in the impact area. This was because these were addressed separately. The rows were the areas of the business and the columns were the business function that was automated. The sample of this table given in Figure 8.6 shows six phases of implementation. The elapsed time for implementation was more than six years. The sequencing was based on the resource level and the availability of a system that could be reused between areas of the bank.

Hetsun Retailing

Hetsun had made a failed effort in several projects. They had developed detailed plans that no one paid any attention to. Project management was not held in good repute. This made the implementation strategy even more important, since political factors dictated that you could go to management with a detailed plan.

Lansing County

In Lansing, operations people were very concerned with how change would be carried out. Rather than develop the improvement implementation strategy without the employees, a different tact was taken. Meetings were held to group the Quick Wins in each area. This was followed by the development of the road map, which was then reviewed in more detail.

Business area	Collections	Customer Service	Payments	Application Processes
Installment Loans	1	2	3	4
Credit card	2	3	4	5
Real estate lending	3	3	4	6
Commercial lending	4	5	5	6

Figure 8.6 General implementation strategy for automation at Kosal Bank business functions.

Other Examples

Two examples in this section deal with e-business. The first example here is a business-to-consumer e-business application for mutual funds. The table is given in Figure 8.7. The areas addressed included IT, processes, and staffing. Costs were not included here since they were not a major factor.

The second example is a petroleum firm that is doing business-to-business e-Business with suppliers. The table is presented in Figure 8.8. Note that the supplier focus exists in both the areas and the impacts. You can also see the approach of focusing on key suppliers where most of the transactions occur.

LESSONS LEARNED

- Start building the improvement implementation plan as you do the work here. This will help you provide management with a date as to when the plan will be presented.
- Try to get widespread participation in departments about the action row that pertains to them. Get employees' views on the Quick Wins and the groupings and sequencing.
- Try not to get into too much detail on each small step of implementation. Point out that each step will be addressed during implementation.

Area	Today	Quick Hits	E-Business
IT	Batch processing; manual	Web-enabled system to get online information for customers	Web application; e-payment
Process	Dispersed; 3 departments	Streamlined from 3 to 2	Down to 1
Staffing	10–12 in processing	5–8 in processing	3–5 in processing
Benefits	Personal service by agents	E-mail; customer satisfaction; staff savings; productivity	Competitive position; fees
Risk	Noncompetitive	Lack of acceptance	Security; identification
Customer	Basic site; mutual fund information	Online information	Full function; online trading

Figure 8.7 Improvement roadmap for mutual fund e-business.

Processes	Manual	Select ordering, payments; ombudsman	Expand transactions
Suppliers	1,000; 80:20 rule applies; manual	Key suppliers (high transaction volume, willing to do e-business)	More suppliers (get key to pressure the others)
IT	Existing online purchasing systems	Web-enabled system for suppliers	E-payments and more functions
Benefits	—	Cut administrative costs	Further savings; improved coordination
Risk	Productivity needs to rise	Suppliers extend schedule; lack of uniform procedures	Lack of (extranet) participation; security issues (extranet)
What suppliers can do	Existing web page	Key suppliers can do limited functions	Posting; updating

Figure 8.8 Improvement roadmap for procurement in petroleum e-business.

PROBLEMS YOU MIGHT ENCOUNTER

- Be prepared for pressure on dates. The problem in committing to dates now is that you have not performed the analysis of the workload or determined what people are available.
- Management has supported the improvement up to date. Here you may run into resistance, because what lies ahead means spending more money and having to deal with policy and political issues. Be open about this. You do want to know now if management will want to shelve the effort. This is better than getting started and having the improvement effort killed off.

WHAT TO DO NEXT

1. Developing a small improvement implementation strategy table can be an interesting exercise. We have done this with attendees to seminars. It is true that the table is very general given the limited amount of time. However, it does show people what is involved. Use the web page example or some process as the basis for this.

2. With a table defined, you can now consider alternatives. Try out the methods and suggestions provided in this chapter to develop three alternative strategies. You might make one conservative and one aggressive alternative.
3. Consider using the approach in this chapter with standard IT and other business projects. It can provide insight into the interrelationships of the work in the projects.

CHAPTER 9

Develop Your Improvement Implementation Plan

INTRODUCTION

Here you might expect to see some stuff on standard project management. Surprise! You won't. Our experience indicates that many of the standard project management methods that work for construction or engineering partially or totally fail when applied to process improvement or are not applicable at all.

There are some key differences between modern and traditional project management.

- *Project manager and leaders.* In standard project management you have one project leader. This gives authority and accountability but can cause trouble in process improvement. In process improvement the work will go on for some time. You have to deal with many issues that are technical, business, and political. One person cannot do all of this. An individual lacks the range of skills required. Also, a single person will get burned out. If the single project leader departs, then you are left high and dry. This is a single point of failure. Therefore, you should have at least two project leaders. At any one time only one of them is in charge—ensuring accountability.
- *Full time project team members.* Standard project management assumes that people are on the team full time. This is a myth. We have not had this happen in more than 30 years of process improvement work. Employees and managers have their standard work to do. If you take them away from the work totally for an extended period, this may cripple the work in their departments. Instead, you want to have people who will dedicate themselves to specific tasks in the project and

then return to their departments. Also, you may involve some people on a part-time basis.

• *The smartest employees.* You are often told that you need to get the smartest people in departments to work on the project. This has many problems. One political problem is that often their power and position is due to the status quo. They really do not want change. If you bring them on board full time, you may have just brought on board your worse enemy! Here you want junior people who are supportive of change and who have energy. Yes, you will need the senior people later to get the detailed business rules nailed down, but this need is limited.

• *Getting the people.* Conventional wisdom says that you should get the people ahead of time at the start of the project. This theory is seriously flawed. How do you know who to get if you don't know the people? And because you haven't started work, you really don't know what skills you will need.

• *Number of people.* In traditional projects, you are told to get a few people and rely on them. This does make it simpler, but you are less likely to be successful. The other employees will resent the attention given to the few. Moreover, you will be relying on only a few people for all of this knowledge. Look at the major religions of the world. They did not grow by preaching to two people in a tent. Many people were preached to. When you change a process, you are implementing a culture change as well as process change. In order to succeed you want to involve as many people as you can.

• *Issues management.* This is critical in process improvement. Technical, business, political, and cultural issues abound. After all, you are implementing change—that dreaded word.

• *The critical path.* Traditional methods focus on the critical path. This is the longest path in the project such that if anything in the path is delayed, the project is delayed—rather simplistic. Who is to say that the tasks with the problems are where you need to spend time on this path? After all, tasks just wandered onto the critical path due to a mathematical formula, not a management assessment. Instead, you will want to devote attention to tasks that have risk and issues or problems. We say that a task has risk if there are one or more issues behind it.

• *Dividing up the project.* Often, people don't give this step much attention. They just divide the project up by departments. Parcel out the project. That is, indeed, the most direct and obvious. However, in process improvement it gets exactly the wrong result. By organizing process improvement in this way, you merely reinforce organization silos and isolation at a time when you are trying to get people to work together across departments. Here we will use two standards for dividing up a project. The first is risk. We divide out a subproject if it has risk. System integration for IT work is an example. The second is to group transactions that cross departments and assign people from different departments to each group. This promotes your cultural goal of having people work together throughout the organization.

• *Project structure.* Traditional project management stresses the work breakdown structure (WBS). This is a complete list of tasks for the work. There are several problems with this method. In any process improvement project, there are going to surprises—some pleasant and some not so nice. A WBS does not provide flexibility. Second, experience reveals that people become more committed to the work if they define their own tasks with their managers and then update their own work. However, you do need standardization so it makes sense to have high-level standardized tasks. This is part of what will be called a template.

- *Collaboration.* People work together in business processes and in doing work and transactions. Thus, it is important to have people work together in doing tasks in process improvement projects. In fact, this is so important that you should have about 30 to 40% of the tasks jointly assigned (with one person serving as the lead person). Conventional project management typically assigns one person to a task to ensure accountability by way of contrast.
- *Project administration.* Much of project management deals with administration—updating the plan, holding meetings, and so on. The goal of this chapter is to reduce this overhead to an absolute minimum. Any minute that you devote to these things takes time away from real work. We think that you should spend most of your time in dealing with issues and opportunities, in communications, and in doing real work—not in overhead.
- *Lessons learned.* Traditional project management often treats projects individually. At the end of the project there is a passing reference to gathering experience. However, anyone who has worked on real projects can see the futility of this tradition. People have left and the people remaining don't have the will to gather experience. There is the urge to go on to other work. The approach is very different here. You want to gather lessons learned as you go. You want to have department staff members gather their lessons learned to help with their work as a Quick Win. The only way you can improve your process improvement efforts over time is to constantly learn and update your techniques and methods.

OBJECTIVES

Technical Objectives

The technical goal is to develop a workable project plan that can be used for the process improvement work now and can later be updated and expanded with lessons learned to improve your future efforts. You want to have the plan developed by the team so that team members not only understand it, but also support it.

Business Objectives

The business objective is to develop a plan that implements process improvement with the available resources in the time frame that is acceptable to management. Another business objective is to provide for flexibility to accommodate change. The plan also has to address the sharing of resources between the process improvement effort and the other work of the people involved.

Political Objectives

There are many political objectives. Let's start with the team. You want the team members to not only understand the plan and work, but to be excited about doing the work. That is part of project management. You want people to be committed so that when they have a choice between regular work and your process improvement effort, they will select your project.

Now let's turn to the political goals for management. You want to use techniques that keep managers informed, but only involved on a selective basis in terms of issues and opportunities. You certainly don't want to have management constantly involved. Managers don't have time and their overinvolvement will drive you nuts. Another goal is that management has no surprises or a minimal number of surprises.

There are political objectives for the departments involved. The first is to achieve Quick Wins so that the process improvement effort can be justified and supported. A second goal is to ensure that the department is not disrupted by the overinvolvement of employees in the improvement work.

END PRODUCTS

The obvious end product is the overall project plan for improvement implementation. However, it goes beyond this. Here is a list of what you want to have as milestones:

- Subproject plans based on the Quick Wins by phase and the areas of the future process
- A method for addressing issues and collecting lessons learned
- A definition of issues that apply to the implementation and association with the tasks to which they pertain
- A lessons learned database to support further improvement efforts

METHODS

Where to Start

You need to start with as much as you can. Otherwise, you will need to invent everything from scratch. Here is the step-by-step approach for you to follow to minimize overhead:

- Review the issues checklist in Figure 9.1 and determine the issues that are relevant for the project at the start. This will get you focused on issues and problems. This will be a key factor.
- Begin with a project template. Figure 9.2 presents a template for the overall improvement effort covering all of the steps in the book in terms of tasks. Note that the tasks are numbered. You want to do this to make communications about the plan easier. Another benefit is that if a task is seen as political, then you can use the number so that you don't have to say the words of the task in a meeting.

Use a Project Template

For each of the entries in your improvement implementation strategy, you will want to develop a template of 15 to 30 high-level tasks. Next, you will define the appropriate milestones. For each Quick Win, you will generally have the following tasks:

- Organize the employees for the implementation.
- Identify and address any potential problems in the implementation.
- Prepare any materials and review these for the Quick Wins.
- Implement the Quick Wins.
- Monitor the initial work after implementation.
- Measure the results.

The actual implementation tasks will depend on the nature of the change. For the milestones you have the following:

- Quick Win ready for implementation
- Quick Win implemented
- Results of Quick Win measured

The future process tasks are also dependent on the situation. The tasks in Figure 9.2 include some of the most commonly encountered ones for information technology (IT), for example.

Determine the Tools That Will Be Used During Improvement Implementation

In the past, few tools were available to help manage the work except manual methods. This is not the case today. There are a variety of software tools that can help:

- *Project management software.* Various packages are available. Microsoft Project is the most popular. It and others support both single and multiple projects.

Potential Issue	Applicable?	Comments
There are delays in related projects' impact tasks		
No one set up or thought about interfaces between projects		
Too much of the work depends on external organizations that cannot be controlled		
The facilities of the effort are not conducive to work		
Competitors working on similar projects are ahead of the improvement effort		
No one considered the impact of external factors at the start of the improvement effort		
There is no systematic method for identifying and addressing external factors		
There is a lack of effort to obtain decisions		
Decisions are being made in an ad hoc manner and not on an organized basis		
There is a lack of consistent project controls		
Projects receive uneven management attention		
There is no standard for project reporting		
Parts of the organization oppose the improvement effort—a factor not addressed by management		
Actions that follow from decisions are not consistent with the decisions		
No follow-up after decisions		
There is no organized forum to address issues		
Managers are dabbling in the work		
Management doesn't hold the improvement team accountable		
There is significant top management change		
The need for improvement has evaporated		
The project is unsuccessful in competing for resources		
The effort assumes a political life of its own		
There is a lack of review of milestones		
Management is too formal and emphasizes status as opposed to content and issues		
There is no ability or desire to pull all of the projects together for analysis		
There is the lack of an organizational process for issues		
Management overreacts to issues		
Misunderstandings about the project are not corrected		

Figure 9.1 Issues Checklist

Potential Issue	Applicable?	Comments
There is no baseline plan		
Resource consumption is too variable		
Improvement meetings lack focus		
There is a lack of awareness of the history of the project		
The project is being treated too routinely–like past projects		
The improvement plan lacks key resources		
There is too much elapsed time between milestones		
The project gets off to too slow a start		
The plan is not kept up to date		
The tasks are more sequential than was planned		
There is too much detail in the plan		
The purpose of the effort was not clear at the start		
The scope of the improvement effort is expanding		
With expanding scope, there is no budget or resource addition		
People are unwilling to change the structure of the effort		
The plan is not synchronized with the actual results		
No grouping of issues. They are being addressed one at a time		
The project plan is changing too much		
Levels of effort and schedule are consistently underestimated		
There are significant missing tasks		
There are too many detailed tasks		
The wrong tasks are deleted in trying to meet the schedule		
There are an insufficient number and frequency of milestones		
The improvement effort is going on too long		
The plan is too informal		
The plan should be changed significantly, but is not		
The plan is adaptive neither to results or events		
The improvement leader lacks personnel skills		
There is no organized approach for sharing of information		
There is too much bureaucracy in the management of the improvement effort		
People are spread too thin in the work		

Figure 9.1 *(Continued)*

Potential Issue	Applicable?	Comments
There is a failure to delegate to the team		
There is no regular flow of end products from contractors		
Team assignments are being changed too often		
People are kept too long on the team		
The project manager is spending too much time in project administration		
There is miscommunications with line management		
The project manager lacks management skills—no one fills the vacuum		
There is an uneven distribution of skills and knowledge		
Team members put out false information		
The project manager spends too much time away from the improvement effort		
Too many people are doing project management		
There is a failure to learn lessons		
People are spread thin between high priority projects		
The style of the team is affecting the project too much		
Boundaries at the edge of the project (indicating scope) are breaking down		
The anticipated resources fail to appear		
There is a gradual deterioration in management communications		
There are poor communications due to personality clashes		
The subcontractor team may not be managed		
Too much of the improvement effort is being done by one person		
Poor performance by some drag down performance by all		
People are relaxing too much after meeting a key milestone		
Even though a method or tool has proven to be a failure, it is not discarded or replaced		
Different people are employing different, incompatible tools		
Staff members are resistant to the methods or tools		
There is a lack of experts in the methods and tools		
No one took enough time to learn the method or tool		

Figure 9.1 *(Continued)*

Potential Issue	Applicable?	Comments
Bad habits in using tools and methods are not addressed or unlearned		
No new tools were investigated during the improvement effort		
A new tool is inserted in the effort without planning		
No one knows whether the methods and tools in the work are really helping		
There is too much rework in the effort		
No one is seeing how the work is being done to look for improvements		
There are too many unplanned tasks		
The quality of the work is an issue		
The work is hit with compound issues and crises and does not recover		
There is a lack of review of what was done		
There is a lack of information on how to evaluate the work		
One issue dominates the work in the effort, leading to problems with other tasks		
Issues surfacing around the work are not taken to the project leader and team		

Figure 9.1 *(Continued)*

- *Spreadsheets.* This can be an alternative to project management software. You will probably want to use a spreadsheet for analysis of the project data. You can export data from the project management software into the spreadsheet.
- *Graphics software.* This software is used for presentations.
- *Database management software.* This is useful for issues, lessons learned, and tracking events in the work.
- *Collaboration software.* An example is Lotus Notes. It can replace using e-mail and the database management software and it supports collaboration.
- *Electronic mail.* This is a blessing and curse. Many people are overwhelmed with e-mail. Moreover, you cannot structure the content of e-mails so that it can be reused. Therefore, be cautious and give specific guidance for using e-mail.

When considering which tools to employ, consider the following guidelines:

- The software should be in use within the company already. Taking on a new tool when you already have risks in the improvement work is just asking for trouble.

- 1000 Overall process improvement vision, objectives, and plan
    - 1100 Develop improvement vision
    - 1200 Define the improvement objectives
    - 1300 Evaluate process improvement management
        - 1310 Identify issues management approach
        - 1320 Identify management approach
        - 1330 Establish improvement steering committee
        - 1340 Establish the improvement executive committee
        - 1350 Identify project leaders
    - 1400 Identify organization roles
- 2000 Process improvement strategy
    - 2100 Define overall improvement goals
    - 2200 Determine mission of organization
    - 2300 Identify alternative strategies
    - 2400 Perform strategy evaluation
    - 2500 Document/present strategy
    - 2600 Strategy review
- 3000 Improvement implementation approach
    - 3100 Evaluate the vision of the organization
    - 3200 Assess supplier alliances and relations
    - 3300 Assess current marketing and other initiatives
    - 3400 Develop an approach to technology
    - 3500 Perform comparative analysis
    - 3600 Define approach
    - 3700 Review approach
- 4000 Identification of improvement candidate processes
    - 4100 Define core processes
    - 4200 Group processes to include related processes
    - 4300 Create comparison tables for processes
    - 4400 Identify finalists
    - 4500 Evaluate finalist processes
    - 4600 Make final selection
- 5000 Competitive and marketplace assessment
    - 5100 Define ongoing competitive assessment approach
    - 5200 Identify internal resources to participate
    - 5300 Investigate benchmarking
    - 5400 Identify specific sources of information
    - 5500 Define evaluation methods
    - 5600 Collect information
    - 5700 Organize information for long-term use
    - 5800 Perform analysis
    - 5900 Present results of analysis

Figure 9.2 Template for Process Improvement

- 6000 Technology assessment
    - 6100 Evaluate hardware in terms of processes
        - 6110 Suitability and support of processes
        - 6120 Identification of missing hardware components
    - 6200 Network assessment for processes
        - 6210 Internal network capacity and performance
        - 6220 Security available and required
        - 6230 Extranet/intranet requirements
    - 6300 Assess system software
        - 6310 Core operating systems
        - 6320 Database management systems
        - 6330 Desktop systems
        - 6340 Utility software
    - 6400 Test environment
        - 6410 Test hardware
        - 6420 Test network
        - 6430 Test software tools
    - 6500 Development environment
        - 6510 Development hardware
        - 6520 Development network
        - 6530 Development software tools, languages, libraries, environment
    - 6600 Identify alternatives for process support
        - 6610 Hardware
        - 6620 Operating systems
        - 6630 Network software/management/security
        - 6640 Development environment
        - 6650 Test environment
    - 6700 Define technology direction for new architecture
        - 6710 Hardware
        - 6720 Operating systems

- 6730 Network software/management/security
- 6740 Development environment
- 6750 Test environment
- 6760 Interfaces with legacy and existing systems
- 6770 New software
- 6800 Staff IT
- 6900 Develop comparative tables
- 6A00 Document technology assessment
- 7000 Information gathering on current processes
- 7100 Direct observation of processes
- 7200 Identify issues in processes
- 7300 Evaluate interdepartmental interfaces
- 7400 Review process documentation
- 7500 Assess current Web activities

Figure 9.2 *(Continued)*

7600 Perform analysis and develop comparative tables
7700 Determine fit with goals
7800 Document
7900 Review current processes

8000 Definition the future and long-term processes
8100 Generate alternatives for new processes
8200 Assess alternatives
8300 Develop comparative tables
8400 Assess technology requirements for future and long-term processes
8500 Establish staffing requirements for future processes
8600 Compare future and long-term processes with current processes
8700 Documentation
8800 Review new processes

9000 Measurement
9100 Identify areas of risk
9200 Define measurement approach
9300 Evaluate infrastructure/technology/support
9400 Measure current business
9500 Measure business
9600 Measure customers and suppliers

10000 Development of the improvement implementation strategy
10100 Define alternative strategies for processes
10200 Define alternative strategies for technology
10300 Define alternative strategies for organization/policies
10400 Define alternatives for marketing and other units
10500 Conduct assessment of alternatives
10600 Develop overall improvement implementation strategy
10700 Define prototype/pilot activity
10800 Define phases for implementation
10900 Review the improvement implementation strategy

11000 Definition of the improvement implementation plan
11100 Define the improvement implementation plan template
11200 Identify specific implementation issues
11300 Assess the project management process
11400 Determine implementation leaders/team composition
11500 Develop detailed plan and subprojects
11600 Analyze the completed plan

12000 Improvement implementation
12100 Set up hardware
12200 Set up and test network
12300 Assess firewall/extranet/security
12400 Develop environment setup
12500 Test environment setup
12600 Establish development standards

Figure 9.2 *(Continued)*

12700 Set up quality assurance
12800 Install software packages
12900 Establish external links
12A00 Test software for production
12B00 Make changes to current application software
12C00 Assess interfaces between current and new software
12D00 Evaluate integration and testing
12E00 Assess quality assurance and integrated testing
12F00 Assess procedures and training materials
12G00 Evaluate operations procedures
12H00 Evaluate network procedures
12I00 Changes to current processes and workflow
12J00 Address current customers and suppliers
12K00 Conduct test of workflow and processes
13000 Post implementation assessment
13100 Gather lessons learned
13200 Identify unresolved issues
13300 Conduct usability assessment
13400 Conduct performance evaluation
13500 Assess impact on current processes
13600 Assess customer-supplier relationships
13700 Perform cost benefit analysis
13800 Define recommendations for later work
13900 Conduct review

Figure 9.2 *(Continued)*

- If you have to buy additional software licenses, then keep them to a minimum. Some improvement efforts we have seen went "hog wild" on tools and blew out the budget.
- There should be an internal expert for each tool you use.
- You should prepare guidelines for the team to use the tool. Don't just throw the tools out and tell people to use them. This will result in inconsistency and other problems.
- If training in the tools is necessary, first train in the method of how you want the team to use the tools. Then only train in the tools for very limited purposes.
- After the tools and guidelines are disseminated, get people using the tools right away. In that way any problems will surface quickly.

Construct the Improvement Implementation Plan

Well, you have the templates, the team, and the methods and tools. It is time to develop the improvement implementation plan (called "plan" for short). Here are the steps to follow.

- For each action team or area of the strategy, assign template-level tasks to every team member. Instruct team members to define detailed tasks at the level of one to two weeks. In that way the work will not be extensive. For small Quick Wins you can just employ a checklist and avoid the plan. Where possible, assign joint tasks between two people with a specific person in charge of each task.
- Each team member will define their tasks and return them to you for review. Here are some questions to answer when reviewing the tasks. Are the tasks complete? Is the wording of each task clear? Does the work result in end products that you can measure? What will be role of each team member in the task? What issues or potential problems do team members see in the work? How will they do the work? What else are they doing that might interfere with the improvement effort? By asking and getting answers to these questions, you uncover more issues and you get the team members to focus on the details of the work. You are also finding out about potential resource conflicts.
- After the review, the team members can revise their tasks, insert dependencies among tasks, add additional resources, and estimate how long each detailed task will take.
- The team members present the results of their work to the improvement leader for review. In this review, you can probe what is behind their estimates of dates. Typically, they may have padded the time due to unknown factors. Flush these into the open and add them to the list of issues. Try and get them resolved.

The approach has several advantages. First, team members are thinking about the work prior to estimating dates and durations. Second, you are finding out about the abilities and knowledge of the team members as you go. Third, and most important, they are defining their own work. They now have participated in the planning process and are more committed to the work. They also know better what they have to do.

Now you have a plan for each area of the improvement effort. You put it together and analyze it. A commonly encountered problem is that the end date is not acceptable. How do you get the final dates improved? The traditional approach is to go to the critical path or the longest path in the project and try to shorten it. This often does not work. Tasks with issues are seldom on this path. Tasks for which there are no issues and, hence, no risk are difficult to shorten. First, you try to see if more work can be performed in parallel. You may have to break up tasks to get results. This will provide some improvement. Next, you should consider the issues behind the tasks and attempt to address them. If these steps do not work, then you might consider the scope of the Quick Wins and future process.

How do you handle contingencies? You could have people pad their tasks. However, if everyone does this, the timeline of the work could be several years long. Instead, define contingency tasks and add them to the plan at the bottom. Do not link them in with the plan. They are ready for use later. This method shows management what contingencies you have planned for and shows your concerns. Here is a list of contingencies to consider:

- The IT part of the work takes longer than planned.
- Individuals are taken from the improvement team for other work.
- A department cannot support Quick Wins or longer-term improvements due to pressing work (e.g., year-end closing).

When you are about to start work in a given area, review the tasks to be done with the team members involved. At this time you can also review and address issues that require attention.

Turning to the budget, you should define the IT, facilities, and nonlabor costs separate from the plan with input from IT and other groups. You can extract the labor hours from the plan and use the plan as the basis for the budget.

Implement Issues Management

Issues management and communications are probably the key attributes of a good project leader. Issues have to be tracked and analyzed. A project leader does not control the people. They belong to departments. A project leader does not control the budget. However, the project leader can control timing. A project leader is, in a sense, like the director of a movie.

You can determine when issues need to be surfaced to members of the steering and executive committees. You can also set the time and place for discussing issues among the team members. Process improvement is political. Therefore, it's useful to escalate issues informally up to management. Try to get as many issues as possible resolved informally.

Use the issues database to track issues. For any new issue, first determine if it is just an old issue in new clothing. You will likely see the same issues recur again and again across the action teams. Be aware of this tendency and try to resolve issues across the teams.

Management Communications

Let's start at the team level. You have to hold meetings with the team. Often, project leaders hold meetings on a regular basis, say once per week. However, this negates one of your few powers as a project leader—timing. When things are going right, hold the meetings less often. When there are many issues, hold meetings more often, say twice a week. Remember that as a project leader what you do is more important than what you say.

What should the meetings deal with? Don't fall into the trap of status. In a meeting with their peers, few people are willing to admit that they have problems. Instead, gather status prior to the meetings. There are two subjects for the meetings: issues and lessons learned. Experience reveals that you should hold about two meetings of issues to one of lessons learned. In the issues meetings, cover a small number of issues (three to four) and keep the meetings short. Don't try to resolve the issues in the meetings. You can do that informally after the meeting. In many societies this is culturally more acceptable. The lessons learned meetings will be covered next.

Organize Lessons Learned

When you are putting in Quick Wins, you want to gather experience as work is performed. In the project meetings on lessons learned, have experience presented to the group by team members who have completed the work. After talking about what they did, you will want to determine the following:

- To what situations does the lesson learned apply?
- What are the guidelines for applying the lesson learned?
- What results are expected results from using the lesson learned?
- What benefits are expected from the lesson learned?

Lessons learned are very useful in the departments as Quick Wins. You can follow the same approach as you used for the improvement team.

When you begin working on new Quick Wins, you can review the applicable lessons learned. You can discuss with the team members how they can be applied.

Evaluate the Plan

Now suppose that you have the completed plan in front of you, along with the list of issues cross-referenced with the tasks. How do you review what you did? Here are some actions to take:

- Extract the labor hours from the project management software and put the data into the spreadsheet. Calculate the total labor hours. This can help serve as the personnel budget.

- Compare the plan with the improvement implementation strategy. Is there a close correspondence? Does the plan maintain the sequencing of the strategy?
- Look at the human resource demands over time. Is the loading of people onto the tasks uneven? Is it even possible to do the work?
- Consider the impact of the major issues on the plan? Have you developed sufficient contingency tasks?
- Highlight all tasks that have substantial risk. Are the tasks with risk and issues spread out over the time span of the project, or are they concentrated at the end? This could be a major problem. You want to review these issues sooner rather than later.

Update the Improvement Implementation Plan

Let's assume that work has been done on the improvement effort. In the old days the project leader sat in a meeting and asked people about the status of their work. With the notes from the meeting, the project leader returned to his or her desk and updated the plan. The approach has several drawbacks. First, the team members are not committed to the work. They just sit and talk. Second, the approach takes the manager down to the level of a clerk. No wonder people don't want to be project managers!

A better method is to have the team members update their tasks twice a week. They can use paper or do the update using the network. They first mark the tasks that are complete. They fill in tasks later out in the plan. There may now be other tasks that slipped as well as additional tasks. They should add these tasks to the plan in the form of additional tasks with dependencies on the original tasks. They should give their initials and state the reason the tasks were created. Examples of reasons are absence due to illness, other work, unplanned work, new changes, and rework to fix previous work.

The project leader then reviews the updated plan. If the schedule overall has slipped, the manager can determine which employees' input caused the slippage. A visit to the employees can then lead to finding more issues and resolving the problems.

Resource Management

Let's start with resource planning. Your improvement effort requires a number of people. These individuals are doing their normal jobs and may be involved in other projects. Construct the table shown in Figure 9.3. In the first column will be a list of the people on the team. In the figure you see Walter and Mary. In the

	Regular work		Other projects	Improvement effort
Resources				
Walter	100%			20%
Mary	50%	20%	30%	40%

Figure 9.3 Resource analysis for the improvement effort.

first columns, you enter the percentage of their time spent in their regular jobs. In the next set of columns, you enter the percentages that they spend on other projects. In the last column, you enter the percentage that you would like to have them on the improvement effort.

You can immediately see two problems. Walter spends all of his time in the department. He does no other project work. It will probably be easier to get him on the improvement effort than someone, like Mary, who is working on other projects. Mary is split between other projects and regular work almost evenly. You want her on average to be spending two days a week on the improvement effort. This is the most difficult situation. If you cannot find a substitute for Mary, then you are probably going to "arm wrestle" with the managers of the other projects.

This table is useful since it highlights the problem in a simple way that can be addressed by management at the start of the process and at the beginning of each phase. You cannot assume that just because management wants to do process improvement that everyone will drop what they are doing and run to the improvement effort.

In process improvement you have three battles. The first is to get the effort started (Chapter 1). The second is to get funding (Chapter 7). Here you have the third battle—getting the people. If you fail in this last one, you fail overall.

From previous process improvement efforts, it is clear that a major issue that repeatedly surfaces during the work is that of resource conflicts. People on the team get pulled off for other work. Guidelines were given to minimize this. These included going for junior people who are less likely to be pulled off and having several people assigned to a task for backup. However, resource conflicts can still occur. What do you do? A solution is to employ a weekly resource allocation meeting with other managers. In this meeting you can discuss what each critical person worked on during the current week. Then you turn to the next week and jointly determine what each person will be doing next week. You can use the same format as that shown in Figure 9.3 to assist the group. After the meeting, the person to whom each affected team member reports relates the allocation to that person.

Develop the Implementation Plan Scorecard

How did you do? Figure 9.4 gives a checklist to evaluate your performance and that of the team in building the improvement implementation plan.

EXAMPLES

ASC Manufacturing

ASC had many years of project management experience. There were no issues in creating the project plan. The problem was that ASC employed a very traditional approach that used full-time project schedulers as part of a project office. The project schedulers had little knowledge of the work. They called meetings to gather status and then ran back to their offices to update their schedules. In theory, it sounds good. Managers do not have to devote their time to this work. In practice, it was a nightmare. It took days before people received an updated schedule. Moreover, because the project schedulers did not know the work, they made many errors.

As a result, the managers who supervised the work began keeping their own schedules. Everyone above the level of technical employees was keeping his or her own schedule. They would then compare their schedules in additional meetings without the project schedulers. This nightmare process was replaced by the method described in this chapter. The results were immediate. There was one central schedule that people could view and update. Many hours of effort were saved.

Another problem was the allocation of people across the work and the improvement effort. The improvement team managers had worked to get the best people, but those people kept getting taken away. Finally, the improvement team

Element of the Scorecard	Score	Comments
Completeness of the plan		
Consistency with the strategy		
Issues associated with the plan		
Elapsed time in developing the plan		
Collaboration by team members in defining their tasks		
Ability to get resources for the team		
Acceptability of schedule and budget		
Extent of contingency tasks		

Figure 9.4 Scorecard for improvement implementation plan.

was shaken up and replacements were made. The problem and impact were reduced.

Kosal Bank

Kosal Bank had simple project plans. After all, the bank was doing more standard system implementation with process change. Kosal's focus was on issues. All meetings were spent on issues. Kosal's management later realized that people were running into the same issues again and again. After this, lessons learned were instituted as subjects of the meetings.

Hetsun Retailing

No previous project management experience at Hetsun had worked. As noted in the previous chapter, there had been failure after failure. Project plans were created but were at a high level for management. Issues management was started for the processes and then applied to the improvement effort itself.

Lansing County

At Lansing County the IT organization embraced project management. People in IT wanted to control all projects. This worked for small internal IT projects, but failed with larger process improvement efforts in other organizations. For the process improvement effort that has been discussed, the approach was to keep the IT staff members at arm's length from the work. They were provided with summary information. This approach helped prevent interference with the work.

LESSONS LEARNED

- To ensure that your plan is flexible, stick to the templates. In that way, any changes that you make to the detail can fit under the template tasks.
- Set a baseline plan. The baseline plan is what was agreed to when you presented the plan to management.
- Consider having 30 to 40% of the detailed tasks jointly assigned. This has several advantages. First, it provides for backup. Second, you instill more collaboration and joint work.
- If the improvement effort is going to be done over a time of six months or more, it will be difficult, if not impossible to define detailed tasks

more than three months ahead. Therefore, have the team members estimate at the template level for tasks further out than three months. As time progresses, the team members can fill in future detailed tasks so that at any time you have a detailed schedule three months in advance. This produces a more workable schedule that is less subject to change.

PROBLEMS YOU MIGHT ENCOUNTER

- Employees who are working on the team are not using the tools.
- You have inherited an overly detailed, unworkable plan.
- People are not working on the improvement effort.

WHAT TO DO NEXT

1. You should gather several project plans that have been developed and employed in your organization. Review these plans in terms of organization, structure, and the other aspects discussed here.
2. Now consider issues and lessons learned. How are issues handled? Is there an effort to gather lessons learned? Was the outcome of the project evident from the quality of the project plans?

PART IV

Implement and Sustain Process Improvement

CHAPTER 10

Accomplish Quick Wins

INTRODUCTION

As you will note, the structure of this chapter is different from the others. This is because the method for implementing Quick Wins differs for each type and each individual Quick Win. Putting a Quick Win into effect also depends on cultural and political factors active in the organization and region. The organization of the method section is based around types of Quick Wins. Examples are then given and discussed.

Up to this point, you have identified the best Quick Wins. You have organized these around your improvement implementation strategy. The improvement implementation plan then has detailed out the schedule and resources needed. You will be implementing Quick Wins in several waves prior to installing the future business process.

OBJECTIVES

TECHNICAL OBJECTIVES

The major technical objectives is to implement Quick Wins with as little disruption to the business as possible. A related objective is to monitor the processes after the change to ensure that the people involved do not revert back to the way things were prior to implementing the Quick Wins.

Business Objectives

There are several business objectives:

- The existing processes must not be disrupted in the transition. Thus, there must be a way to turn on the processes in one or a few days—not over an extended period.
- The Quick Wins need to be mutually supportive of each other or independent. They cannot be conflicting when put in place.
- The Quick Wins in one phase must build on those in earlier phases.
- It is necessary to implement a regular measurement method for processes and change.

Another objective is that you are paving the way for the future processes.

Political Objectives

The political objectives are basically twofold:

- To implement and measure the Quick Wins in a phase or wave
- To arouse enthusiasm and confidence in the changes

The first objective is important for management. Nothing is worse than changing processes or making other alterations and then not measuring the results. Management may begin to think that the situation has not improved. Thus, tangible measurement by a group outside of the people involved in the Quick Wins is critical for credibility. You can employ the same structure of the improvement steering committee and executive committee to report on the results.

Achieving the second objective will assist in the future process changes by giving people confidence in the approach and effort. As opposed to talk and meetings, they are actually seeing progress. This gives the improvement team support and also spurs on the action teams.

END PRODUCTS

There are end products at three levels. There are individual Quick Wins for a specific process. Moving up, there are all of the Quick Wins for a phase. Finally, there are all Quick Wins installed prior to the future processes. For each there are the following end products:

- Results of the Quick Wins on the processes
- Impact of the Quick Wins on the employees and morale

- Lessons learned from the transitions in processes and their use in improving successive transitions

As you can see, there are both tangible end products relating to results and intangible ones relating to attitude and morale.

METHODS

WHERE TO START

Sit down with an action team that addressed some of the processes that will be affected by Quick Wins and carry out the following steps:

- Review the process and transactions as they exist.
- Jointly determine how the transition and change will occur.
- Identify what supporting and preparatory actions are necessary prior to transition.
- Discuss any potential negative events and factors that may arise during the transition.
- Present the transition to the employees with the supervisor and someone from human resources present.
- Carry out the transition.
- Monitor the first week of the process after the transition. Make any other adjustments.
- After the transition, gather the action team, supervisor, and employees to review the transition. Gather and document lessons learned in how the transition was accomplished to better prepare for the next phase of change.

If you follow these steps, things will proceed smoothly. The approach gives flexibility in adjusting to unforeseen circumstances.

You will want to involve the internal audit or finance staff members to help in measuring the benefits of the Quick Wins. First, make them aware of the work performed in Chapters 5 to 7. They will have to buy into the impact of problems and the measurement of the processes prior to any changes. Second, provide them with information during the transition. Determine jointly when it is appropriate to measure the benefits and impact of the Quick Wins. It will be very difficult to separate out the exact benefits attributable to a single Quick Win. Therefore, it is recommended that you carry out the measurement step at the process or department level.

There is benefit in involving the human resources department in the Quick Win installation. Job duties may change. Human resources needs to be aware of changes to help guide in the transition to the modified processes. As an example of the situations that could arise, let's suppose that a Quick Win is put in place that eliminates most of the work that an employee does. The department in question and the human resources department must be quick in reassigning the person to other work areas.

Examples: Process Related

Quick Win: Automate a Small Manual Accounting Process

An example might be the reconciliation of credit card transactions. This may have been a manual process when the volume of work was low. In order to support more volume later, this will have to be automated. Otherwise, it will prove to be a bottleneck when the future process is installed.

This Quick Win can be accomplished by buying a software package or constructing a database application. The software is first set up. Then the data are converted over to the small system. After training and going live, you can then archive the manual methods.

Quick Win: Gather and Share Lessons Learned in Departments

We have employed this Quick Win many times. It is one of the most obvious and least frequently carried out Quick Wins. Why? People do not see the need. They may see each piece of work as unique. What do you capture for a lesson learned?

- The title of the lesson learned
- The situations to which it applies
- Who uses the lesson learned
- How staff members are to employ the lesson learned
- The anticipated result in using the lesson learned
- How the lesson learned will be disseminated to the staff

In addition, you want to describe how the lesson learned will be updated through experience. For the short term it is useful to have a group meeting to discuss how staff members applied it and what they got out of it. This will lead to further guidelines.

Lessons learned basically improve the way work is performed. If lessons learned are gathered early, they can be employed as business rules in future processes.

Quick Win: Make People Aware of How Much Time Is Spent in Exceptions

Chapter 5 identified some exceptions. However, there was no systematic effort to be complete. Here we make an effort to identify more exceptions. For each, the following information can be gathered:

- Situation to which the exception applies
- Procedures followed for the exception
- Cause of the exception
- Frequency of the exception
- Volume of work that fits the exception conditions

What do you do with this information? The first thing to do is to make everyone aware of the extent of the exceptions and the problems that they cause. Second, you can get down to the root cause of the exceptions and, perhaps, address these causes in later Quick Wins. Third, for the exceptions that remain, you can make sure they are performed more systematically.

Quick Win: Review and Modernize Existing Procedures

In Chapter 5, you probably uncovered a number of problems in the detailed procedures. In some instances there will be no formal procedures. This is an opportunity to make the procedures more formal and structured, and it directly paves the way for the future processes.

Quick Win: Identify and Analyze as Many Shadow Systems as You Can

Here the action teams scour departments for shadow systems. There are several benefits in doing this. A shadow system needs to be out in the open before it can be eliminated. People have to become aware of the problems generated by the system. A second benefit is that additional requirements are uncovered that might be addressed in the future system.

How do you find these systems? Train in the work procedures. A second way is through observation. Watch how less frequent transactions are performed. Find the people who perform the unusual work and work alongside them.

You can classify your information by answering the following questions:

- What is the purpose of the shadow system?
- Who uses the shadow system?
- How is it employed in the work?
- If the shadow system did not exist, what would the employees do?
- What is the impact of the shadow system on the overall process?
- How often is the shadow system used?
- What was the reason that the shadow system was created?
- What did employees do prior to the shadow system being built?
- What are the business rules behind the shadow system?

What actions are possible with a shadow system? These three are common:

- You can replace it by the future system. This means that the shadow system was performing necessary and legitimate work.
- You can eliminate it by changing the policies or business rules. This means that the shadow system filled in for a faulty policy or rule.
- You can suppress the shadow systems. However, if there was a perceived need, it will just come back.

Examples: Communications and Coordination

Quick Win: Employ Off-the-Shelf Collaborative Software

In our past work, we have used this approach effectively a number of times. In remote construction and engineering sites, collaboration tools were installed. In another instance, off-the-shelf software was used to support sales tracking. A third case involved deploying a new online system in more than 20 countries in parallel. Lotus Notes is the prime example of these tools. Goldmine and similar software are other examples.

Here are some of the benefits of this Quick Win:

- People can share information on issues and problems and how these were overcome.
- Employees can structure information in a way that is impossible with e-mail. In e-mail, one person sends another a note about three problems. The recipient responds on one of the problems and raises two others. There are now four outstanding problems. However, there is no structure in place to deal with them.
- Project plans, reporting on the work and progress, and other aspects of project management can be more easily supported through collaborative tools.
- A more standardized and systematic approach to work situations in distributed locations can be supported through these tools.

Quick Win: Become Aware of Department Communications

In everyday work situations, most employees go to their jobs and perform the work. They might talk during breaks, but often these conversations are personal and not business related. Then they go home. Management never gives them the opportunity to just sit around a table and talk. What is there to talk about?

- The employees can identify potential further improvements.
- They can share information on how they do their work.

- They can tell each other about interesting situations that they encountered. This is useful even if it is to share a joke about a situation with a customer or a supplier.

Quick Win: Make Employees Aware of What Is Going on in Other Departments

As with the previous Quick Win, people are largely unaware of what is going on in other departments. In many cases, work in other departments depends on or impacts their work. Employees may wonder why they are having to do their work in a particular manner. They begin to understand, however, when supervisors from other departments make brief presentations to them. The supervisor might also point out problems that he or she encounters. The employees might address some of these problems as they go about their own work. It is another opportunity to improve the process. Why not do all of this in Chapter 5? There is often insufficient time.

Quick Win: Perform Review Training in a Department

In many organizations people learned how to do their work on the job, or they received formal training years ago. If you perform refresher training, it can reinforce good practices in doing the work. Alternatively, it may surface opportunities to improve the training and make it more relevant.

Examples: Management

Quick Win: Carry out a Contest Internally to Collect Information on the Web about Competitors

After doing this several times, we have found that it is one of the more useful Quick Wins to apply. One benefit is that employees and their families become more aware of the industry and competitive firms. They may have an idea for a better approach than what is done in their own work environment. They may also see that the methods that they employ and what they do is similar to methods used in many other firms. A second benefit is that this may trigger more ideas for improvement. The ideas people come up with tend to be good ones because people are very familiar with their work habits.

Quick Win: Try to Identify and Address Instances in Departments Where There Is a Single Point of Failure

In some departments, there is only one person that performs a specific, critical task. This is a single point of failure. Another example occurs where there is a lack of formal procedures and training in a unit. The unit then is a single point of failure.

Why is this important? Because anything adverse that happens in this work situation will tend to directly impact the productivity and output of the department. Work may back up when a person is gone on vacation or sick leave. Another impact is that work is slowed down because others know the absent person's work is backed up. Why hurry the work if it will just wait at the next step?

Single point of failure situations should be addressed in the future process. It is more difficult to address them with one or two Quick Wins.

Quick Win: Develop Specific Mission Statements for Departments

Mission statements were shown to be useful in Chapter 2 and used in later chapters. At the department level, the mission statement helps make employees aware of the importance and role of their work in the overall company. The mission statement is good for motivation. Many managers assume that people know these things—a bad assumption.

How do you develop a mission statement? First, get the company mission statement and break it down into specific thoughts. Second, identify key processes and work that are performed in the department. With these items in hand, create a table in which the company mission statement elements are rows and the processes are columns. The entry in the table is the goal for the process in terms of the specific mission element. Now look for common elements in the table to create the mission.

Quick Win: Implement Scorecards for Two to Three Processes in Each Department

Most of the preceding chapters have offered improvement tables and scorecards. Here you would employ the process scorecard to assess a small number of processes in the department. You should do this for processes that will not be addressed by Quick Wins or the future process.

There are a number of benefits to implementing the scorecard and showing it to the employees. First, you get discussion going as to what score the process should get. Second, people will examine why the score value turned out the way it did. This may give rise to more improvement ideas.

Quick Win: Have Supervisors Meet to Share Lessons Learned on Handling Employee Issues

Supervisors are in the same situation as the employees they manage. There is just no time for getting together to share ideas. If you provide a forum and opportunity for them to get together, then they can see where they can improve their own work with the ideas and experience of other supervisors.

In order to start these meetings, you can pose some questions, such as these:

- What is the most interesting thing you have seen?
- What are your most common employee issues?
- If you could have money to improve something, what would you spend it on?

Quick Win: Identify and Review All Ongoing Projects

At any given time in medium and larger organizations, there are often many small and midsize projects going on. Most of these are not systems related. Rather, they pertain to producing information for management, handling a regulatory requirement, and similar situations.

Getting all of these projects out on the table can provide a view of what activities beyond normal work might interfere with the implementation of the future processes. You can also see which ones really are not needed and which should be combined, since most projects are started individually at different times. You might also find opportunities where some of these can be eliminated or killed off.

Quick Win: Carry out Instant Downsizing

This is what many firms do today. Downsizing often turns into a disaster. Incentive programs are offered to get the most senior people out. What happens is that the individuals who can find work elsewhere leave and the remainder try to hang on. This means that the good people leave and a firm may be left with a group of gnomes and trolls—not good for the processes.

Another approach is to eliminate an entire organization or business unit. Here the company may not have thought through the longer-term impacts of the action. There now may be holes in processes that the remaining business units have to fill on an emergency basis. This won't be cheap.

Instead of instant downsizing, a better approach is to consider the business processes when making these types of changes. The company should ask and answer the question, "What processes do we want to give up?"

Quick Win: Implement Uneven Downsizing

This pertains to international firms. Different countries have unique laws that limit the options for downsizing. There may be longer notice periods for shutting down a facility. Employees may have to be paid over a much longer period. This is the case in many countries in Europe. In such cases, the international firm resorts to making disproportionate cuts in countries and locations in which there are no such rules. On the one hand, this can be viewed as unfair. However, what it means ultimately is that the countries without the rules are more productive and efficienct since they can scale up or down in response to market conditions.

There is another, more basic point. Have you ever worked or observed people working in a plant that should have been shut down but was not? There is little motivation. The employees know that the only reason they are there is because of government intervention. Equipment is maintained and there is some production, but there is no drive for improvement. In some cases, we have observed that new shadow systems and procedures emerged to slow the work down!

Examples: Customer and Supplier Related

Quick Win: Gather Information on the Top Five Questions Asked about the Top Fifty Products

This pertains to customer service and sales. Often in a retailing firm, the people who are doing the selling do not know as much about the products as the employees who are handling customer questions and complaints. Why does this occur? The people work in separate departments and may not run into each other on a daily basis. Also, people tend to stick to their own work. Management does not provide an opportunity for these employees to get together because it fears such interaction would reduce productivity.

Sharing information about the products and services of the firm is very useful. First, you can improve how sales are done. Second, customer service employees can be more effective in dealing with customers. Third, the information about the product can be used on web sites and brochures or catalogs, thereby generating more sales and more satisfied customers. These steps help to focus employees on the customers and their needs.

Quick Win: See How Suppliers Are Doing Their Work

Several departments deal with the same suppliers every day. A useful Quick Win is to visit the supplier and see how the supplier does the work. This gives employees of the supplier firms a chance to surface their gripes and problems with your firm. You can then work with them to solve these problems. After all, they are common problems and negatively impact the productivity and attitudes on both sides. Have the employees of your firm express their problems with the suppliers.

These meetings also serve a useful political purpose. They show the supplier that you really care about the relationship. This may not get you a better price for the merchandise, but you will likely get better service. After all, how many other customers are doing this? Another related benefit is that the employees become more supplier focused.

Unusual Examples

These two seemingly simple and stupid Quick Wins indicate that nothing legitimate can be left off of the table when you are considering Quick Wins.

Quick Win: Provide More Handcuffs

In implementing a new point-of-sale system in a drugstore chain, a number of interviews were conducted with store employees and management. In each case, questions were asked relating to the problems in their work situation and with the existing computer system.

At one store, the store manager was asked, "If you could have money to fix something, what would it be?" We thought that the response would relate to the computer system, inventory, or the store layout. It did not. She said, "I need and have asked for another pair of handcuffs, but I never get them." Taken aback we asked why she wanted these. She indicated that when people come in to shoplift and steal merchandise, they come in pairs. She only had one set of handcuffs and so could only apprehend one person. She had requested the second set of handcuffs for several years to no effect.

Needless to say, the second pair of handcuffs was delivered in a week. It was keyed to the same lock as the original pair. She was extremely grateful. The improvement team had demonstrated tangible results that she had not been able to obtain. In the following month, her employees apprehended three additional suspects for shoplifting. Stock losses dropped as thieves moved on to easier targets. Store managers as a result gave the improvement effort much more support.

Quick Win: Install Tables and Chairs

In a manufacturing operation, work goes on across three shifts when times are good. Many manufacturing firms are located in depressed areas where land and facilities are inexpensive. In one such plant, there had been a security problem at night. The outside and covered areas where employees ate their meals in the middle of the night were locked off. Employees had no place to eat except on the floor. They had complained about this for months, but management did not do anything because they did not see it as a problem. This impacted both morale and productivity. Within two weeks after seeing the problem and the impact, chairs and tables were installed. Morale climbed as did support for the improvement effort.

Develop the Quick Win Scorecard

As noted in previous chapters, it is useful to develop and employ a scorecard to assess the effectiveness of your work. Figure 10.1 shows the first implementation scorecard.

EXAMPLES

ASC Manufacturing

At ASC there was initially little management desire to implement Quick Wins, although they were viewed as nice to have. The managers' initial focus was on the future process. This all changed when they saw the schedule for the software for the future process. Suddenly, they realized that if they were going to get any benefits in the next four to five months, they would have to implement some changes. For economic reasons, management kicked off the Quick Wins. The managers' attitude changed again when they observed the change in attitude among employees. More support and attention were given to Quick Wins. This took some of the pressure off of the software development team as well.

Element of the ScoreCard	Score	Comments
Economic results obtained from the Quick Wins		
Ease of implementing later Quick Wins based on the experiences witht the earlier ones		
Success rate in implementing Quick Wins		
Degree of resistance encountered		
Surprises uncovered in the work		
Number of additional opportunities that surfaced		
Support and participation of employees		
Extent to which action teams shared information		
Schedule of implementation		

Figure 10.1 Scorecard for Quick Win implementation.

Kosal Bank

Some of the Quick Wins were discussed in Chapter 7. Here we point out that employees became very involved in improvements as the initial Quick Wins were implemented. They saw a chance for improvement. This was especially true in several units where there had been high staff turnover. With the Quick Wins and the prospect for further improvements with the future process, turnover declined and morale improved.

Hetsun Retailing

There were two founds of Quick Wins. The first yielded more than $1.4 million in savings. The second round yielded a saving that was five times that figure and included revenue increases. Here are some of the additional benefits that were evident:

- There was lower staff turnover.
- More people requested over time.
- At breaks people discussed their work.
- More suggestions for improvement came in.
- A number of small projects in departments were canceled due to lack of benefits.
- Customer response was almost immediate in terms of letters of praise instead of complaints.

Lansing County

In any government agency, it is sometimes hard to imagine a Quick Win. The words "quick" and "bureaucracy" do not seem to go together. Here they did. Employees in the operations area belong to a union. The union had for some time complained that the problems of its members had not been addressed. This had led to a very difficult union bargaining. As the Quick Wins were implemented, the members pressed their union officials to support even more changes. Working conditions improved. The next union contract was negotiated in record time. All of this occurred prior to the implementation of the future process!

LESSONS LEARNED

- Keep track of the experiences and lessons learned gained from the implementation of the Quick Wins. This will not only help you to deal with the same departments later, it will also give you more confidence in working with other departments.

- After doing some Quick Hits with one action team, have the team members discuss their experiences with members of other action teams. This helps to gain sympathy for the changes and for the problems encountered.
- Managers will often want to know how things are going. In addition to a status or progress report, consider showing them an example of a transaction in a department. By giving an example you are making the change tangible. So that they are not under the impression that this is easy, also point out the issues and problems that you encountered as well as the surprises you uncovered.

PROBLEMS YOU MIGHT ENCOUNTER

- Even with all of the planning and analysis in the world, you are still likely to encounter surprises. A shadow system suddenly shows up. An exception turns out to be critical. These are just two examples. The improvement team and the action teams need to be on the lookout for these things at all times. Some may not even surface until after the transition to the changed procedures.
- You have addressed the subject of resistance to change. Often, it is academic to people until the change and transition begin. The team should assume that there will be some resistance to change even after all of the planning and preparation.

WHAT TO DO NEXT

1. Select specific transactions from a process and see what Quick Wins you can implement. Then try to move up to the process level.
2. Try to determine in this analysis what barriers you will likely encounter in implementing change.

CHAPTER 11

Implement Improved Processes

INTRODUCTION

In the preceding chapter, you grouped and implemented the Quick Wins. The improvement implementation effort tends to be sequential and parallel at the same time. You will begin with a prototype of the new process in terms of automation. You will then initiate a pilot test of the new process using the prototype. During the pilot, you will uncover gaps and holes that have to be addressed. You will find where the existing policies must be changed or replaced.

There are major areas during implementation where there are risks and issues. These include data conversion, integration and testing, training, and business acceptance of the new process. Data conversion refers to the establishment of information so that the new processes can be operational. Integration is significant because it is here that many software errors occur. For that reason testing is linked to integration. Training includes both the new systems and new processes. Training also includes reinforcing the understanding of the problems with the processes both prior to Quick Wins and after the last round of Quick Wins. In this chapter, we will spend time examining each of these areas. Note that since every implementation is somewhat unique, the material in this chapter will serve as guidelines.

OBJECTIVES

Technical Objectives

The main objective is to get the new processes to operate effectively. Measurement and coordination of the operational processes follow in the next chapter. More specific objectives include the following:

- Develop and test the new systems and the new transactions through prototyping and piloting so that policy changes can be identified and the system design and process can be validated.
- Address key areas that have been found to be risky and involve many issues. These include data conversion, integration and testing, training, and the cut-over of the new processes.

Business Objectives

One business objective is to ensure that the work is not disrupted by the implementation of new processes. Another business goal is to get the processes installed as well as the new system.

Political Objectives

A cultural objective is to have people in different departments collaborate on the implementation effort. The hope and expectation are that the collaboration will continue. Another goal is to put managers and supervisors on notice that management takes process improvement seriously and seeks to expand it.

END PRODUCTS

The end products depend on the specific situation. Here is the list of milestones that apply to many improvement efforts:

- Prototype of the new systems for the processes
- Pilot test and evaluation of the new processes using the prototype
- Implementation of infrastructure, network, facilities, hardware, and system software
- New or modified policies to govern the new processes
- Data conversion from the older systems, manual files, and other sources to prepare for the new processes
- Integration and testing of the new systems

- Documentation and training
- Cut-over to the new processes

If you are implementing a software package, then the prototype is replaced with installation and setup of the software package. The pilot then tests the package and the new processes.

METHODS

WHERE TO START

The key is to organize parallel efforts. You cannot afford to do this sequentially: there is too much pressure. The information technology (IT) work is defined well enough. Turning to the business units, more effort is required. You have the action teams in place. Look at the end products in the previous list. Here is how the action teams and employees and supervisors can be involved:

- *Pilot test and evaluation.* Testing the prototype with new transactions will reveal some surprises and surface some issues that will have to be addressed.
- *Training in the use of the network and infrastructure, if this is required.* An example might be the deployment of a new e-mail system that was made possible due to the infrastructure changes.
- *Development and review of policy changes and new policies.* The impact and determination of how the policies would work are important here.
- *Data conversion support.* This is critical since technical people and the improvement project team may not know about the data in detail.
- *Testing of the complete new systems and processes.* Testing involves more than system testing. The processes must be tested.
- *Issues management.* Business employees and supervisors are involved in resolving issues related to exceptions, workarounds, and shadow systems.
- *Operations procedures and training materials.* Operations procedures include the user procedures for the new systems.
- *Training of employees in the new processes.* This includes the new systems.
- *Support for the cut-over to the new processes.* This includes dealing with any remaining problems that surface and the killing off of the old processes.

PROTOTYPING AND PILOTING

At one time the concept of a prototype involved a model of the system. It was not functional. The prototype provided an idea of what the user interfaces and data elements would be like. It could not process information. Prototypes today lack interfaces to other systems and also do not include all transactions, reports, or business rules. However, they do more often include the user interface, transaction processing, and processes for updating and viewing the work. This makes for a more realistic prototype. It also reflects the fact that many systems are Internet browser–based so that the user interface or graphical user interface (GUI) is basically determined.

Successive prototypes are developed and delivered for review based on employee feedback. The prototype will never go into production, but if carefully built and designed, it can become part of the final production system. The prototype is the basis for further development of the software and for evaluation of the process during the pilot.

Prototypes are often designed around the workflow of transactions. Associate a status code with each workflow step. Identify what triggers a change in the step and, hence the status. You can get this information from the work in Chapter 7. Workflow is very important since it allows for workflow tracking and can generate productivity statistics. It can also assists in e-business transactions.

Another concept is that of transaction queuing. Transactions are entered by customers online using the Internet. While some of the processing is performed in near real time, the transactions build up for handling orders, complaints, requests, and so on. This is a queue. You can apply various business rules to determine which items in the queue are processed first, second, and so forth.

In Chapter 7 you also defined the job functions that will perform the work in the transactions. By aggregating these functions, you can determine the menu options for employees, supervisors, customers, and suppliers.

The prototype also includes the databases to be used to support the transactions. These include the master databases such as the customer database. There are also temporary databases such as an order database. In addition, there are lookup and reference tables such as the zip code lookup to get cities and states or provinces.

After the first prototype has been developed, it is ready for review. This often includes the menus and navigation among screens. Several screens for entry, search, and update are also available. In the review you can concentration on the following elements:

- The look and feel of the user interface
- Coverage of the job functions by the menus for each job
- Coverage of transactions through the menu options
- The layout of the data entry screens

A wide range of employees should review this information after the action team and supervisors have provided their input.

Successive prototypes add features and functions. Additional transactions may be added. You will generate a list of problems and questions during the prototyping and reviews. For each you have to decide on what to do. You have the following options:

- Implement an item in a later prototype.
- Set the item aside for further evaluation.
- Include the item during later development.
- Drop the item as being out of the scope.

At the end of the prototyping stage, collect the remaining items and issues. Add these to the issues database as part of issues management (discussed later). When are you finished with the prototyping? The technical answer is when the system is ready for the piloting.

The pilot test is the assessment of the new process with the last prototype. Your goals are the following:

- To validate that the new transactions can be supported by the new systems to some extent.
- To verify that the new system is relatively easy to use and will likely generate fewer problems than the old system.

It is useful to consider the pilot in two parts. The first part is an in-depth assessment. To make this assessment, you will assemble a small group of people to review the processing of each transaction or type of work. You may also consider how the system can address exceptions. If the new system is to replace a shadow system, then you can see that it will. In this first part of the pilot, you are interested in the following:

- Comparing the old, new, and latest Quick Win versions of the processes in terms of ease of use and potential performance
- Detecting ambiguities and gaps in the new process
- Uncovering and testing new policies to govern the work
- Examining controls to see if errors are detected
- Assessing the effectiveness of measurement

To proceed, gather a small group and have the members identify transactions and situations. Gather data from the real work world. Divide up the situations. Now show the group how to do the testing and how to write down the results. Remember to keep telling the group members that the more problems they find, the better. Reluctance to point out problems at this stage can lead to severe problems later. As with the prototype effort, you will be identifying issues and problems for tracking and handling.

The second part of the pilot is to expose the process and system to a wide range of employees in departments. Politically this will generate grassroots support for improvement effort. Many improvement efforts have failed because of lack of involvement here. During the second part, you hope to complete the following tasks:

- Uncover the learning curve for the system and process.
- Determine ease of use with a broad cross-section of people.
- Solicit additional suggestions for improvement.
- Gather support for the changes to come.

A specific technique that can be effective and that was employed at two of the organizations given as examples is that of the continuous demonstration. Here you demonstrate the process and system at lunchtime to groups of employees. You can offer soft drinks and cookies. After the demonstration, have members of the audience "come down" and try the process out. This shows how easy it is to use and takes the mystery of what has been going on out of the picture. Circulate a small form for the employees to complete. The form should contain the following information:

- Date
- Position of the person (not the person's name)
- Number of years in the company
- Rating of the following items on a scale of 1 to 5 (1 is low; 5 is very high):
  –Ease of learning
  –Ease of use
  –Completeness in terms of the interface
  –Improvement over the current system
- What did you like best?
- What did you dislike most?
- What would make the process and system better?
- What do you like best about the current process?

Tabulate the responses over time. You can also modify the questionnaire based on feedback. In the Lansing County example, more than 120 employees were involved. The number was more than 180 for ASC Manufacturing.

The continuous demonstration approach has several benefits:

- Widespread feedback is obtained.
- Resistance to the new process later is lessened.
- Bottom up support for the new process is generated.
- Many lessons learned are collected and can be used in development.
- More exceptions are addressed.

Remember that you are dealing with a process group, so you will want to consider the cross-impact among processes and work in the group. You will also review the interfaces with other work in departments that will not be touched. It is these interfaces that can destroy productivity gains generated by the new processes. This is another reason to involve employees who handle different processes—not just the ones that you are changing.

Policy Changes

Policies play a very important role in process involvement. You can often make minor changes during the Quick Wins, but the major changes occur during implementation of the new process. The new systems make possible the policy changes. These changes and the automation then facilitate modifications to procedures. During the pilot work when you are testing out the new transactions is a good time to determine the new, detailed policies for the new processes.

A policy change often is triggered by an issue. When the issue surfaces, define a potential new policy and compare it to the old ones. Try to develop precise terminology for the new policy. Then you can address the following questions:

- What is the impact of the policy on the processes?
- What is the effect on control and audit? Can the work be reviewed to see if the policy is being followed?
- What is the scope of the new policy? To what work will it apply?
- How should the organization transition from the old to the new policy?
- How easy is it to circumvent the new policy?
- What is the impact of the policy on exceptions, workarounds, and shadow systems?

You will have to document the new policy. Include the following items:

- New policy description
- Corresponding old policies
- Differences between the old and the new
- Impacts if the new policy is not implemented
- Benefits of the new policy if implemented
- Transition approach to the new policy
- Impact of the policy on other processes

Infrastructure, Hardware, Network, and Software

Work on the infrastructure probably started during the Quick Wins due to the lead time. Infrastructure is where you spend the money in process improvement. Technology infrastructure includes telephones, cabling, communications and networks, hardware, software, and IT-related staffing. This activity is sequential, so you want to get started soon. The lessons learned from the pilot will also provide input on the quantity and type of hardware in some cases. Nontechnology infrastructure includes modifications to buildings and facilities, office layout, office improvements, and furnishings.

A general guideline here is that when this step is completed, make sure that employees see the progress and the positive impacts on their working lives. Nothing is more frustrating than seeing things torn up, and then when stability returns there is no change.

At Lansing County, the infrastructure was basically in place. Some additional hardware was required. For Hetsun Retailing, new PCs had to be acquired. The network had to be upgraded. Similar changes were acquired at ASC Manufacturing. The new processes at Kosal Bank required new hardware and software.

Data Conversion

In most cases, moving to a new system and process means converting information from the old process and system into the new. Data typically exist in the automated system, manual files, and local files on PCs. Lack of attention to conversion has been one of the main reasons why process improvement can fail. Consider what might be wrong with the current information:

- Information is faulty because there has been little update validation.
- The data have not been used and so is suspect.
- Data in the manual and automated parts of the process are inconsistent.
- Information may originate in other processes and systems. Changes could have been made here, contaminating the data.
- Data in the current system are not in the proper format.
- Missing data must be found, captured, and entered.
- The new process may require new data elements, requiring more data capture.
- History data are in a different format than that of the active master file of data.

Even with modern technology, data conversion may not be simple. Extensive manual labor may be required. In conversion, quality and completeness of information are crucial. To get error-free transactions requires accuracy at each transaction step. Otherwise, additional manual labor may be required to find the problems and make corrections. Productivity suffers as a result. Poor data may cause process deterioration.

Where do you start? Begin with the new processes. Determine requirements for conversion based on sources of information, quality, and processing characteristics. You will want to sample the existing data to determine the problems.

Considering the following factors when making data conversion decisions:

- *Timing.* If you transfer or reconstruct data too early, you will have to accommodate changes to the data prior to the new process being implemented. This may require the construction of an entirely new, though temporary, updating process and system. If you are too late in this effort, the conversion will delay implementation of the new process.
- *Quality of the information.* Test and sample data to determine the quality and to identify what data are missing.
- *Location of the data.* Where is the information? Where is the best information—in which system or manual file?
- *When to start the effort.* Since data conversion can be time-consuming, you probably want to begin early.

You will probably have to capture data from manual files to set up the new system and process. Here are some guidelines:

- Arrange files into batches and put them in a central location. If the information in the files changes, the division into batches helps make the search for specific files easier.
- Carry out a quick survey of each batch. Handle the usable files first. Separate out problem files for more experienced people.
- Consider scanning and optical character recognition (OCR) for data capture. However, manual entry will still be necessary in most cases.

You still might not have all of the data elements or information that you require. Carry out trade-offs between the value of having the data on one hand, and the effort of capture on the other. Determine the minimum extent and amount of information that are required (not nice to have) to cut over to the new processes.

Each situation is unique. Data conversion was not much of a problem at ASC Manufacturing. There was a moderate amount of conversion at Lansing County. Kosal Bank required substantial manual data entry from handwritten files. For Het-sun Retailing, very little manual information had to be converted.

Integration and Testing

Integration can be a nightmare. Some of the problems from the IT side include the following:

- The new systems and existing interfacing systems may not have matching data elements. Translation and reformatting may be necessary.
- Different systems may process data at different times, raising compatibility issues.

- One or more of the existing systems may be very old (legacy systems) and their data may be of questionable validity.

Each of these factors and others may require not only more programming, but intense and detailed coordination among programmers. There could be a legacy system where no existing programmer has done substantial work. There may be a lack of knowledge. Examples of these problems occur in e-business where the new software must interface with existing accounting, order processing, or purchasing software.

There are a number of factors to consider in creating and testing an interface, including the following:

- Verification of the proper timing of the interface
- Exact data elements to be passed, including formats and order
- Method to be employed in carrying out the interface
- Confirmation that the data has been received
- Approach for correcting errors that are detected
- Implementation of a rollback approach so that databases are not contaminated
- Recovery if the interface fails

There are also process interfaces that must be addressed. You should answer questions such as the following:

- How will the interface be tracked?
- How will balancing and checking occur?
- How will errors be handled?
- How will information be pulled from both processes?

Factors that can impact multiple processes include the consistency of the procedures followed in doing the work function, policies that impact multiple processes, and the infrastructure shared by multiple processes.

There is also integration among the parts of the systems and processes. For example, in e-business you might acquire a number of software products that require integration. Several types of integration can be distinguished:

- *Loose connection.* An example is the employee and the computer system. This is a manual connection and simpler to test.
- *Physical connection.* The activities are performed by the same people even though they are part of different processes.
- *Tight connection.* The activities are totally interdependent.
- *Functional connection.* The activities perform related functions, but are not totally dependent on each other.

You can scratch your head for days trying to split hairs on a specific set of processes. Here you want to be aware that the degree of integration and type of connection affects the extent of testing required. The more integrated the processes are, the more difficult it is to do complete testing. In planning for testing, take advantage of loose connections between parts of the process and systems to reduce the amount of testing.

A number of different types of testing are required:

- *System testing.* The system is testing in its entirety. The focus is on automation.
- *Process testing.* Here the overall process is tested. This includes the system, but the focus is on the overall process, including any manual parts.
- *Transaction testing.* A transaction is taken through the entire process and system.
- *Performance testing.* In performance testing, the system and process are submitted to high workloads. You can determine throughput (the volume of work per unit time) and response time (the time it takes to process a piece of work).
- *Error handling.* Here the methods for handling errors and exception conditions are tested.
- *Acceptance testing.* This is testing by the employees and supervisors (or selected customers or suppliers) for approval.

You can compare the results with the existing processes and systems.

How do you prepare for testing? You have to determine what data to use in testing. You can generate simulated transactions. However, more often people extract production data and modify them for use in testing. The next decision is to define how testing will be done. Test scripts can be generated. These are test cases where there is a defined input and output. This allows for verification that the testing worked.

Testing can be a big effort. Many people have not been trained in testing and quality assurance. This was evident in the downfall of many web sites. Test plans are essential to ensure that everyone is aware of the scope and extent of testing needed. Testing has to be organized and monitored. Test results have to be reviewed. Any errors may surface issues that have to be addressed in either the systems, the processes, or both.

THE ORGANIZATION AND THE PROCESSES

This is another interface area. You want to consider how the organization will work with the new processes. Don't think that you can easily reorganize in the midst of implementation. It will be too difficult and disruptive. That can come later. In considering the interface, answer the following questions:

- How will the organization supervise the new processes and manage the work?
- How will the organization deal with exceptions and problems that arise?

- How will the organization address policy issues and undertake policy enforcement?

You can also start building lists of job duties. Organization issues will be discussed in the next chapter.

Now move up to management and consider these questions:

- How will management control the processes and departments?
- What information will be presented to management?
- What method will be used for reporting problems?
- How will exception management reporting be done?
- What information do the new processes and systems provide for errors, productivity, performance, cost, and staffing?

A tip here is to establish a "strawman" or candidate model of the reporting process and methods. This can then be reviewed with simulated data to get feedback.

Training and Documentation

There are several categories of documentation. First, there is the documentation related to the implementation and IT. This includes the following elements:

- Issues definition, analysis, and decisions
- Integration and test plans
- Test cases, suites, and testing results
- Program documentation for interfaces
- Backup and recovery procedures
- Computer system operations procedures

Then there is the documentation that will serve the process and support the cut-over in the future:

- Policies governing the work.
- Operations procedures for department employees. The procedures for using the systems are included here so that there is one source that is consistent.
- Training materials. This includes orientation training as well as in-depth training for new and existing employees.

Here are some guidelines for the documentation:

- Operations procedures and training materials can be developed using successive levels of detailed outlines. A useful method for documenting a transaction is to employ tables similar to those presented in Chapters 5 to 7. A sample format is shown in Figure 11.1. Note the columns on business rules and lessons learned. The business rules are the conditions

Transaction: ____________________________

Purpose: ________________________________

Management expectations for the work: ________________________________

Step	Who	What	Business rules	Lessons learned

Figure 11.1 Sample format for operations procedures.

that govern the detailed work. If they are automated, then this column can be blank. The lessons learned can be those gathered during Quick Wins. Much of the future value of the procedures will be in the lessons learned that serve as guidelines for the work.

- Parcel out transactions and get a number of people working on these in parallel. Then you can have the employees review each other's work for testing.
- In reviewing the documentation, find out if it is understandable. Also, determine if it is complete. Define how exceptions will be addressed.
- Develop training materials for specific transactions as you construct the operations procedures. This will save time.
- Use the work in Chapter 7 as a start on the operations procedures.

Let's now move to training. Train a small group of employees first. Use a member of the action team as the initial trainer. This will provide a great deal of feedback. First, you will be able to evaluate the effectiveness of the trainers. Second, you can assess the clarity, completeness, and consistency of the documentation. Third, you can find gaps and holes in the procedures to make later improvements. Continue to refine the training as you go.

In terms of the training itself, always provide an introduction overview. The overview should review the old process and procedures as well as issues and problems. This overview will reinforce why the improvement was necessary. Next, give a high-level presentation of the new process. You are now prepared to plunge into the detail. After the group has been trained, the people should go to an area where they can work with sample transactions and data. This work can be monitored closely. Immediate work will reinforce the training. It will sink in better. After this training, the employees can be put to work in the new process.

Issues Management

Issues and issues management have been discussed throughout the book. The tracking of issues and the issues database were covered. Use these same methods and tools to deal with implementation issues. It will be important to track open implementation issues. Also, you will want to carefully document and identify how each issue was addressed. Otherwise, the same issue may recur.

Cut-Over

Over the years, several methods have been used for doing the cut-over to the new processes. In the past, a common method was to do it in parallel—that is, the new and old processes were performed in parallel. This method, for the most part, is not possible today. There is neither the available resources nor time to do this. Instead, we tend to turn on the new processes and turn off the old, which is not as traumatic as it seems with preparation. A third option is to expand a pilot implementation in one area or location. This is also a commonly employed overall approach. In each location, you would then do a one-time turnover.

With the turnover completed, some critical tasks still remain. First, you have to monitor the new processes closely. You will want to take measurements. Second, you want to have the old process archived and destroyed—that is, the old documentation and files should be archived. Forms from the old process can be destroyed. It is important politically that this action be visible to show people that there is no going back. The only road is the road ahead.

Develop the Process Implementation Scorecard

Figure 11.2 gives the scorecard for implementation. Note that, as in other scorecards, many of the elements are subjective.

Element of the Scorecard	Score	Comments
Degree of interaction between IT and business employees		
Completeness, adequacy of testing		
Degree of success in data conversion		
Quality and results of training		
Review results for documentation		
Results of integration		
Smoothness of cut-over		
How issues were handled during implementation		
Outstanding issues after cut-over		

Figure 11.2 Process implementation scorecard.

EXAMPLES

ASC Manufacturing

Overall, the implementation of the new process went smoothly. More than 30 significant issues had to be addressed. During the pilot, for example, a number of policy issues surfaced.

Kosal Bank

Kosal Bank experienced a rough first improvement effort. Many of the people had never done this before. It was successful, but it took longer than planned. The critical success factor for the future was that extensive lessons learned were gathered and used in later improvement efforts.

Hetsun Retailing

The critical success factor in implementation at Hetsun was the involvement of the action teams. This provided consistency and continuity for changes. Issues that surfaced during implementation were rapidly addressed as a result.

Lansing County

At Lansing, the improvement implementations were very successful in operations. However, there was less success in other departments. Due to political infighting and other culture factors, no one from the operations team was allowed to help other departments. They had to learn on their own.

LESSONS LEARNED

Here are some basic guidelines for prototyping. Agree on the frequency of prototypes in advance and stick to it. Make sure that the employee role in the review is carefully defined:

- The larger the time gap between prototypes, the greater the chance that employees and action team members will become disconnected from the work.

- The fewer the number of prototypes, the less feedback you will receive. However, the feedback will be received on the basis of diminishing returns.
- Issuing versions of the prototypes too frequently will mean that many items remain open and that there is little difference between successive versions. The employees will not be impressed and may become turned off.

Some lessons learned from infrastructure include the following:

- Some items such as communications and cabling may be long lead items due to permits and other factors. Start them as early as possible.
- Establish communications capacity that can support three to five years of growth. Of course, the other technology may change in this time.
- Delay hardware purchases as long as you can since these items get cheaper and better in performance over time. Consult Internet auction sites for routine equipment to get some good buys (subject to warranties, inspection, etc.).

Here are some additional lessons learned:

- Even though everyone is busy, keep on the lookout for additional improvement opportunities. Sometimes you can use the pressure of implementation to get some good changes in place. This will be easier to do than incorporating Quick Wins.
- Be sensitive to differences in style and culture when you are doing this work. It is possible that such items can be addressed in the prototype and pilot. One of the biggest failures that we have observed internationally is the lack of testing and implementation work in regional locations. Many people feel that if it works at headquarters, it will work anywhere. This is a false assumption.
- After you have completed the initial testing and employees have gotten the feel and knowledge of how to test the prototypes, leave them alone to let them test on their own.
- As soon as the operations procedures are started, try to divide up the work so that more employees are involved. This will get the work done faster and give you more to review.

PROBLEMS YOU MIGHT ENCOUNTER

- A major problem is that so much attention and effort are spent on the systems side of the work that the final effort of the process implementation may be forgotten. We call this "midinstallation paralysis." It is more common than you think. This is why you must keep attention on the processes.

- Prototypes can be seductive. Departments may push to use the prototype in production. This is, obviously, not possible, but it is triggered by appearances. Requirements may start changing, leading to an unstable prototype—unless there is strong management control. Management may put pressure on you to speed up development of the improvement process based on the speed of the prototype development—another thing to be headed off.

WHAT TO DO NEXT

Obviously, you cannot implement some changes in a short time. Therefore, we suggest that you search the literature for examples and case studies on how implementation faired. Write down the critical success factors and the factors that contributed to failure.

CHAPTER 12

Measure and Maintain Improvement Momentum

INTRODUCTION

At this point, the processes are functional. In many projects this would almost be the end of the story. There would be a post-implementation review and everyone would then "ride off into the sunset." This is also how many improvement and Six Sigma books end. In the real world, process improvement is not like that. There are many things to do. Here is a list to get you started thinking:

- The processes must be measured to ensure that the benefits were attained.
- The processes must be maintained.
- The processes must be monitored to detect signs of reversion to the old process or general deterioration.
- Issues and problems are likely to remain after implementation. New ones crop up. These must be addressed.
- The process improvement effort must move on to other processes.
- There are usually opportunities to enhance the processes further based on the lessons learned.

It is widely thought that a business process is stable and does not deteriorate over time. Such an idea could not be further from the truth. In fact, most of the time the process deteriorates far faster than the IT systems and technology supporting the process.

How can a process deteriorate? Here are some examples that we have observed. Note, however, that each situation is somewhat unique and depends on the work, the people, and other factors involved.

- *Change in another, related process.* Someone makes a change in a process that impacts the one that was changed. The supervisors and employees respond to the situation under time pressure by improvising exceptions. Then the same type of transaction starts appearing more frequently. Now the exception is becoming a rule.
- *Loss of staff.* The people that you trained and spent so much time with leave due to many factors through attrition. New employees enter the department to perform the work. It seems that the supervisor was caught up in emergencies and fighting fires, so no training was performed. The supervisor asked the new people if they had done similar work before. The new employees, not wanting to look dumb or lose their jobs, respond in the affirmative. The supervisor then tells the employees to get to work. Training will be done later. However, "later" never comes. The ingredients for deterioration set in.
- *New business rules.* A new business rule or condition arises due to internal or external factors. The information technology (IT) group will not be able to handle this in the system for several months. The department is forced to implement a workaround. When the system is modified, the workaround never seems to go away.
- *Management change.* There is a change in supervisors or managers. The new managers want to put their mark on the department. They came from company X which did things "right." The manager now directs the employees to implement this approach in the department. The intimidated employees follow instructions.
- *Emergence of substantial new technology.* The technology changes in a significant way, making the systems supporting the "new" processes obsolete. There will be pressure to replace the old technology. This could impact the processes.
- *Organizational response.* There are several potential factors involving organizations. One is that the organization has not changed to fit with the new process. Tension rises. Something has to give. Often, the organization does not change and the processes revert back. Another case is that the organization changes, but not correctly. The new process may be performed in a worse way.
- *New employees.* There is a new employee, often a trainee or intern, who volunteers to develop a small system for a department. A new shadow system is in the making. Watch out.

These changes are not new phenomena. If you read articles from the 1920s that discuss work measurement and industrial engineering, you can detect many instances where positive improvements were undone and the process returned to its original state.

How do you prevent deterioration? There is no surefire way since processes depend on people and organizations. However, if you automate the process, then it becomes standardized. It may be less flexible and harder to change, but the transactions will suffer less deterioration. E-business, when all of the hype dies out, has as its goal the automation of processes. It is one of the best hopes for preventing deterioration. Second to this is an internal online system that performs many functions, including tracking, and eliminates paper.

Unfortunately, in most cases no one budgeted for the activities that have been listed here, so a marketing effort needs to be done.

OBJECTIVES

Technical Objectives

The technical objectives include the following:

- Synchronize IT system changes with the processes.
- Measure the processes to detect any deterioration.
- Pursue additional opportunities for improvement.
- Provide and organize for monitoring processes.
- Ensure that changes in organizations, surrounding processes, and other factors are planned and organized so as to maintain the integrity of the new processes.

Business Objectives

One business objective is to keep up support for the new processes. Another is to get and stay behind additional improvements. Hopefully, the initial improvement effort will encourage others. Be careful here. People get burned out. Next, you can have chaos if you implement many simultaneous changes. A leading maker of clothing found this out when it initiated more than 60 reengineering projects at once. The company went bankrupt.

Another business objective is to now analyze the organization structure and roles. The purpose is to get the organization better aligned to the new processes so that the work is more effectively performed.

Political Objectives

A major political objective is to show how important processes are and that the organization should become more process focused. This happens and is noted in e-business successes. Companies that have implemented working e-business have found that they become more process focused and less organization centered.

Another political goal is to serve notice that there is a need for improvement and that the methods that were employed can be sustained and expanded to other areas.

METHODS

Where to Start

You want to begin the activities in this chapter prior to completing all of the improvements. To ensure the integrity and prevent the deterioration of the processes, a process coordinator should be appointed. This person can be involved in the last stages of the change implementation. A process steering committee to review what is going on in the processes should also be established before work is completed. In that way, both gain experience and knowledge of the changes.

As part of the work measurements of the processes in the group are taken and provided to management. You also want the process coordinator and members of the process steering committee to be involved here so that they can observe the measurement process at work.

Process Coordination

Process coordination is an important aspect of work operations that is overlooked often. We are not talking here about the day-to-day supervision of people. There are full-time managers to do this. What we are talking about is tracking and monitoring processes in groups to ensure that process performance is maintained and that deterioration is prevented.

To handle process coordination, you need to establish two different roles. First, you should have a coordinator who can organize efforts to fix problems and gather and analyze performance statistics. Second, you want to have a group of people who meet once a month to review their assigned processes. This will be called the process steering committee. The process steering committee gets more people involved in maintaining the process.

Process Coordinator

Critical processes are so important to the business that organizations should create the role of process coordinator. This is not the same as the improvement team leaders. This is a role that should rotate to involve a number of different employees. Experience indicates that the person should be appointed for one year. There should be a three-month overlap for the new person and the current one. The position is part time so that the employee can still perform most of her or his regular work.

The duties of the process coordinator include the following:

- Collect information on processes and issue process report cards to management.
- Coordinate the process steering committee (discussed later).
- Track changes made or requested of the processes.
- Meet with IT, human resources, and other groups to get advance warning of any changes that might impact the processes.
- Design and recommend specific changes and improvements to processes to the steering committee and then to management.
- Alert management to potential process problems.
- Coordinate the implementation of enhancements to processes.
- Coordinate the training of new staff in the business processes.
- Collect information from departing employees.
- Support computer system changes that affect the processes.

Notice that some of the items get into the running of a department—especially the coordinating of changes and the monitoring of employee's being trained and departing.

You could say that the department supervisors should perform this work. This approach will likely fail. They are too busy in their daily work. Moreover, many of the processes that were selected for improvement crossed departments. The supervisors may also have vested interests in the organization and may not feel loyalty to the processes. They may be people oriented, rather than process oriented.

There are two levels of coordination in business departments. The first is the day-to-day coordination of the process work. This is the supervisor's job. The second entails a wider view across departments and policies. That is why we indicated that the process coordinator role should rotate on an annual basis among different people.

Select a junior person who is interested in the work and in technology. The person should also have been involved in the work for some months. Such a person is more interested in change than senior employees and less tied to the old methods and ways. After being appointed, the person can be trained quickly in other departments so that he or she gets an overall view of the business processes.

It may be possible to have this person help IT in the coordination of changes to the systems that support the process group. Here are some of the duties involved:

- Identify and formulate system change requests.
- Work with IT to develop requirements for the changes.
- Identify benefits of the changes.
- Coordinate testing of the changes to the systems.
- Support the documentation of the changes in the operations procedures and training materials.

- Validate the benefits of the changes.
- Determine any additional impacts of the system changes.

The existence of the process coordinator does not reduce departmental or supervisory responsibilities for the processes. They are still accountable for the work. The department managers and supervisors should be responsible for measuring the coordinator's performances.

The Process Steering Committee

People have too much to do already—not another committee. The process steering committee is a positive committee that meets monthly and addresses the following:

- Identifies new opportunities for further improvement to the processes in the group.
- Determines if there are any changes or challenges to the integrity of the processes.
- Uncovers lessons learned and experiences that could be useful in doing the work.

The process coordinator can organize the work of the steering committee. The members of the steering committee are composed of both employees and supervisors from departments that are affected or involved in the processes of the group. There is no need to have upper level managers involved as members due to the detailed nature of the work. By forming a process steering committee, a company is showing its employees how much it values the work and wants to sustain the changes and improvements that were made.

The process steering committee members should be overlapping. To get things started, some should be appointed for six months and others for a year. New members can be brought on board for a year.

Who do you select for the committee? Our approach has been to pick individuals who were involved in the process improvement effort and who demonstrated interest in the process change. The people selected should also be outgoing and aware of what is going on in their departments beyond their immediate duties.

Maintain and Enhance the Processes

Maintaining a process means that you take action to ensure that the process continues to meet its goals and requirements. Maintenance refers to being able to accommodate situations with the process—without creating more exceptions and shadow systems. Examples of maintenance include taking steps to reduce response time or errors to acceptable levels while continuing to provide the same information to customers.

A process may be forced to respond to new pressures and requirements. Examples were noted in the introduction. Meeting new requirements or attaining new performance levels are referred to as process enhancements. These may entail substantial changes. An enhancement may apply to a process, organization, or infrastructure change. Examples of process enhancements are as follows:

- Dealing with additional, unplanned volume with the same process
- Processing a new type of transaction or work
- Obtaining additional staff performance statistics from the process
- Supporting a new service or product line

Enhancements can be either proactive or reactive. In a reactive mode the changed requirement or need appears unannounced. The department employees are forced to respond it to. There is no time for planning. A new exception or workaround is suddenly created.

An example of a reactive enhancement occurred recently in an e-business. Marketing decided to offer a new promotion on the Web. If you ordered $XXX worth of merchandise at one time, you received a 10% discount as well as free shipping. People started to place orders. No one told the IT group or customer service. The result was chaos. With the current software, each order was limited to one shipping destination. A customer could not order a number of items with multiple ship-to locations in the same order. The system would not handle it; customer service could not handle it. Marketing had to create a temporary group to back into the numbers to make it work.

It is clear that reactive enhancements and change are damaging to processes. In a proactive enhancement, people sit down on a regular basis and review what is going on in the process. Then they decide what changes to make—trying to make sure not to damage the process. Because e-business crosses many departments, more process changes tend to be made proactively. A basic observation is that as processes become more automated, they become more structured. Hence, such changes tend to be proactive. In order to get something in quickly without waiting for the systems to change, people are forced into manual workarounds and exceptions.

The mixture of maintenance, reactive enhancement, and proactive enhancement reveals a lot about the state of the processes as well as the organization's attitude toward their processes. Consider these additional observations:

- In general, the more proactive the enhancements are, the more the people in the organization and management care about the processes and realize their importance.
- A lack of maintenance and enhancement activity does not indicate success. Processes are based on human behavior, so they tend to change. If

there is no activity, this may mean that the process is of little importance and should be eliminated or merged with another process.

- If there are many reactive enhancements and excessive maintenance, then the process is undergoing substantial deterioration. Wait a year or so and you probably won't be able to recognize it.

Conduct Measurements

The process scorecard for the old process can be used again. It is given in Figure 12.1. Now you can also go back to the transactions. Here you want to make sure that the steps in the future processes as defined in Chapter 7 have been carried out. It is useful to employ similar tables to that in Chapter 7. Figure 12.2 can be used for critical and frequent transactions. This table is useful because it reveals what additional changes were required. The reasons and any other explanations can be entered in the last column. You will need these tables to get at the benefits (column 5). This will be discussed in the next section.

Now you can roll up the transactions in a summary table. Here you can summarize the benefits as well as identify outstanding issues. Take a look at Figure 12.3 and you can see a column for issues.

When you implement change, you are almost certain to generate new issues. Examples are additional automation changes and organizational problems. You also

Factor	Score	Comments
Number of people involved		
Turnover of business staff		
Number of exceptions		
Number of shadow systems		
Quality of training materials		
Quality of procedures		
Extent of current automation		
Existence of measurements		
Cost of doing the process		
Impact of leaving the process alone		
Number and severity of issues		
Impact of systems and technology on the process		
Extent of potential opportunities		

Figure 12.1 Scorecard for the new business process.

Process: ____________________________

Transaction: ______________________________

Step	Who	What	Difference from planned future process	Benefits	Comments and observations

Figure 12.2 Sample analysis table for the new transaction.

may find that some of the process issues that were uncovered in Chapter 5 were not addressed. Use the table in Figure 12.4 to record information about the remaining issues. Notice the third column. It refers to type. By type we mean the category. Examples include systems, procedures, policies, other processes, organization, and staffing.

You can also now summarize the benefits and the issues in a new table that appears in Figure 12.5. How do you perform these measurements and gather information on the outstanding issues? One way is to start by interviewing improvement and action team members. Get them together in a meeting and discuss the processes. This should trigger an effort to gather data through measurement and direct observation of the work.

Now that you have gathered the information about the outstanding issues, the question arises as to what you are going to do with this data. One thing to do is to develop a spider chart to show the number of issues by type. You could compare this chart with one you create for the old process. An example of such a chart is shown in Figure 12.6. This particular chart was created for an insurance company after one of its key processes was improved. In this diagram, you can see that there were many problems in the original process. While the new process addressed the technical and procedural issues, substantial policy, organizational, and management issues still remain. You can now see one way to use this information—as pressure to get the organization to deal with its problems.

A process report card is another way to evaluate a business process. There

Process: ____________________________

Transaction	Volume	Frequency	Benefits	Issues

Figure 12.3 Summary transaction analysis for the new process.

Process: ______________________________

Issue	Description	Type	Affected transactions	Impacts	Comments

Figure 12.4 Outstanding issues relating to a process.

should be one report card on each process in the group and one for the overall group. You can employ the following grading system:

A— The process is acceptable. No immediate improvements are needed.
B— The process works, but enhancements to the process are possible.
C— The process works, but the benefits are not being achieved.
D— The process does not work for all of the transactions or all of the time. There may be, for example, a high error rate.
F— A substantial number of the transactions are failing to meet acceptable standards.

With the grading system defined, you can consider grade components. A list is given in Figure 12.7.

Validate the Benefits

You can now assemble the charts and tables and start reviewing the benefits. The obvious measure of benefit is cost. Here you can compare the total cost of the new and original process. Make sure that you can adjust for the level of work as this could have changed during the improvement effort. Also, pay attention to the mixture of transactions. The work could have gotten easier or more difficult. Use the row headings of the tables to identify any other changes that could have affected costs.

There may also be revenue benefits. By automating and changing processes, you might be able to offer new products and services. You might reach new mar-

Process	Benefits	Issues	Comments

Figure 12.5 Summary of benefits and issues.

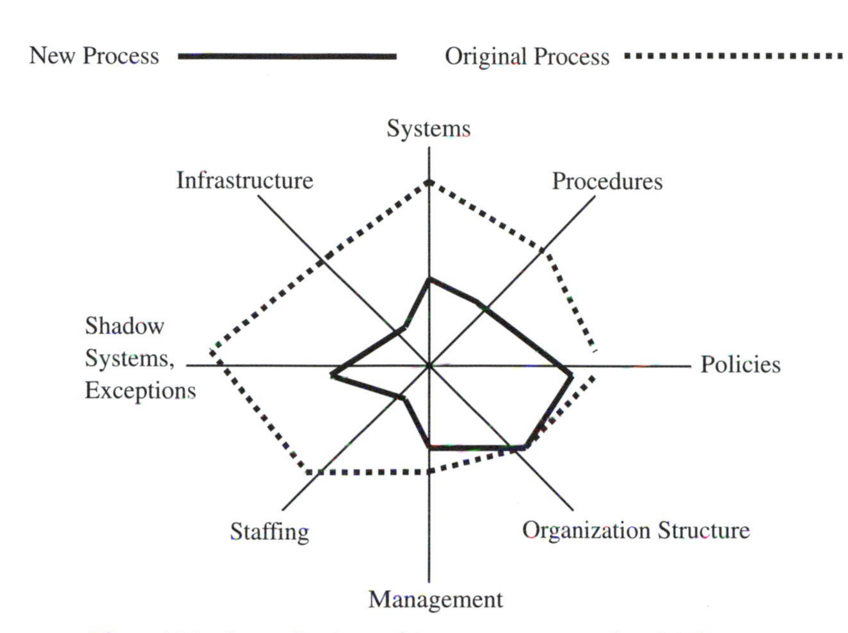

Figure 12.6 Outstanding issues of the new process versus the original process.

kets. For example, we helped a family in Asia get into e-business to sell their furniture. Over 40% of their sales come through this channel.

There are intangible benefits to be considered too. Wherever possible, these intangibles should be translated into tangible benefits. Here are some examples:

- The new process and systems are easier to use. This means that it should take less time to do the work. Productivity should be higher. The rate of errors should be lower. Training of new employees should take less time.
- Information is easier to find and more accessible. This should reduce response time for customers. The number of complaints should drop off. It should take less time to do a transaction.
- We know more about the process in terms of performance statistics. Automated systems produce statistics on the length of time for a transaction and the volume of work by transaction, and they tabulate reported results and data by employees. This makes management easier and supervision less subjective. Less subjectivity and more objectivity should improve morale and reduce turnover of staff.

Suppose that you did all of this measurement and found that the benefits were not up to expectations. Let's also assume that you developed the original estimates of benefits and still feel that they are valid. What do you do? Identify

Grade Component	Grade	Comments
Organization		
Employees		
Systems		
Technology		
Infrastructure, facilities		
Error rates		
Volume of work		
Response time		
Throughput		
Cost per transaction		
Overall process cost		

Validate the Benefits (H3)

Figure 12.7 Grading components for process evaluation.

the reasons why the benefits were not achieved. Structure these into the table in Figure 12.8.

Organizational Changes

This book does not pretend to be about organizational design, and you will note that we have treated organizational change as beyond the scope of the chapters so far. Now it is time to say a few words about organizational change. In most cases, we prefer to change the organization only after process improvements have been completed and measured. It is only after the change that you observe the fit between the organization and the business processes.

How do you start? Begin at the bottom where the work is performed. For each person doing the work in every category, develop a job description. Write down the duties that they perform in the processes. Then summarize these in terms of high-level work objectives. You can also determine standards based on the meas-

Issue	Type	Description	Impacts	Potential actions	Comments

Organization Changes (H3)

Figure 12.8 Issues that affect benefits.

urements that are now possible with the new processes and their supporting systems. This information can now be employed in several ways. First, you can make a side-by-side comparison of what the old job descriptions were. This can be very revealing in that you hope and anticipate that the work will be more interesting and less routine. If it is the same, then you have to ask where the benefits are. A second application is to now develop job descriptions for the supervisors.

Work with the human resource group when you do this activity. Provide human resources with your analysis and make some specific suggestions. Remember that your job is not in human resources. You are just providing input. To further impel them to take actions with respect to the organization, prepare another table of impacts (see Figure 12.9).

Now suppose that human resources develops a plan of action. Volunteer to help in the implementation of the new organization. Do this on the basis of trying to protect the processes and work from disruption. A suggested approach for change is to stabilize the supervisory and employee positions first. Then the process is protected. Middle management changes can now be implemented. Here the concept of the new organizational structure was top down, but the implementation was a combination of high-level changes in management and adjustments in the detailed work.

Address Process Deterioration

Process deterioration will almost inevitably set in regardless of your best efforts. How can you address this in a proactive way consistent with limited resources? A key lesson learned here is not to respond to every little small sign or event. This will drive you and the employees around you crazy.

An alternative approach is to consider what has happened in the processes within the group on a regular basis. How frequently you do this depends on the stability of the processes; a rough rule of thumb is every two to three months. You will, as the process coordinator, gather the process steering committee together and review what is happening. You can then do analysis and develop specific actions.

As with the previous sections, tables can be useful here. Figure 12.10 is one such table. It gives the impact of the issue, potential action, and guidelines for change.

Position	Old tasks	New tasks	Impact if not addressed	Comments

Figure 12.9 Potential impacts without organizational adjustment.

Problem	Source/cause	Impact	Action	Approach

Figure 12.10 Process deterioration table.

The process steering committee and the supervisors and employees can now be presented with the table for their reaction. You want to ensure that when you take action that it is sufficient and that you are making many small changes that drive people nuts. This is termed a batch mode of addressing deterioration. If you were to address each small sign, it would be termed online. Note that for some signs, you will not take action because there is insufficient information. It is better to wait in such conditions.

Enhance the Processes

Process enhancements due to proactive ideas should also be carried out in a batch mode. Here you will follow a mini-version of the improvement effort. Here are some of the key steps:

- Ensure that the employees and supervisors realize the need for change and the problems if the work persists as is.
- Use a collaborative approach in implementing changes and improvements.
- Make sure that you have taken measurements before the change and that you do the same afterward.
- Give credit to the employees and the process steering committee. Here the steering committee replaces the action team, in effect.

Expand the Process Improvement Effort

Let's be positive and assume that your improvement effort worked. Where do you go next? Should you stop? There are powerful forces that come into play for the urge to stop. People are tired. Some may be burnt out. In some e-business efforts, the work is so demanding that you are compelled to stop and catch your breath. E-business in such situations is often best done in waves.

If you follow the approach in continuous improvement and some Six Sigma implementations, then you continue to move ahead on improvements.

This is probably the best approach, since you have momentum and success behind you.

Now that you have decided to consider continuing, where do you go next? Investigate the following alternative approaches:

- Consider surrounding processes. These are typically additional processes in the same departments. This is good because the people are used to change. It is bad in that they may be burned out. These processes may not be the ones that yield the most benefits. After all, they were dropped in the final selection.
- Go back to the work that you did on process selection in Chapter 4. Consider the other finalist group.
- Redo the analysis introduced in Chapter 4 and do a new evaluation.
- Find a high-level manager who supports the improvement effort and wants to do it in his or her organization. This manager will give you a lot of support. This is, perhaps, the path of least resistance.

Develop the Process Coordination Scorecard

As with the other chapters, you can employ the scorecard in Figure 12.11 to assess your efforts. Another scorecard can be created and employed for recurring use. An example is given in Figure 12.12.

Element of the Scorecard	Score	Comments
Appointment of a process coordinator		
Establishment of the process steering committee		
Measurements of the new processes		
Validation of benefits		
Detection of process deterioration		
Degree of coordination of changes with IT		
Department awareness of processes		
Enthusiasm to keep the integrity of the processes		
Awareness of processes by management		

Figure 12.11 Scorecard for process coordination.

Element of the Scorecard	Score	Comments
Extent of deterioration		
Extent of activity of the process steering committee		
Number and extent of outstanding issues in the processes		
Degree to which there is management interest in process improvement and processes		
Number of suggestions for improvements by employees		

Figure 12.12 Scorecard for recurring process analysis.

EXAMPLES

ASC Manufacturing

The implementation of improvement came first in the integration and assembly area. This was originally selected due to complexity. Management felt that if this could be improved, then simpler areas could be addressed. This was also the area of greatest business need. After this was done, the results were presented to management. This was a great success.

The next area was selected. Upper management decided not to make pronouncements of future areas. Instead, a meeting was held in a large conference hall. At the meeting the improvement implementation team summarized the approach and results to managers and technical staff from all divisions. This was then followed by detailed presentations from members of the action teams that participated in the first improvement effort. This was a great success. People's fears were allayed since the individuals making presentations were their peers and not the implementation team. As a result of the meeting and further discussions, two additional areas were identified as the next areas for improvement. The improvement work continued on like this through all divisions.

A middle-level manager was appointed as the improvement coordinator. The improvement steering committee was composed of upper-level managers from the major divisions. The steering committee made the final decision as to which areas to move to—with the input of the employees.

Kosal Bank

A very high level manager had served as the champion for the initial improvement efforts, which lasted a number of years. Other bank managers then took notice of the results and approached the improvement implementation team with further opportunities. This resulted in a reprioritization of the remaining work in the first area. The manager of the IT area served as the temporary improvement coordinator to establish a working relationship. The steering committee was made up of senior managers from major bank units.

HETSUN RETAILING

There had been previous failed efforts where something had worked and then there was no follow-up. This was such a concern that the financial controller volunteered to act as the improvement coordinator for further work. While the systems work was completed in the first phase of the improvement effort, there were many left over Quick Wins and changes. This consumed the improvement effort for the next year.

LANSING COUNTY

In the Lansing County organization, the operations manager who oversaw and directed the initial improvement effort continued the work into other areas. This was successful because of the continuity provided. However, you will find very few managers who would do this.

LESSONS LEARNED

- Take care to make sure that the processes continue to be measured after success. Many often want to stop after benefits have been measured. If you stop measurement, how will you detect deterioration in processes?
- Do not declare victory too soon. This is a problem in many efforts where some success is achieved in Quick Hits and then the systems are installed. The pressure for more process change can wane if people think that the effort is over.
- Try to keep the action teams together informally. You might want to hold some lunch meetings where they review what is going on. This way you can find out if they are working together with each other well.

PROBLEMS YOU MIGHT ENCOUNTER

- A new manager enters the scene. What do you do? You have seen that new managers may start making changes to create processes they were used to in their old jobs. Go to any new manager and bring him or her up to date with the improvement effort. Indicate the monitoring approach that has been put into action. If the new manager still wants to leave an imprint on the organization, feed him or her the outstanding, unresolved issues. This will give the new manager something to chew over.

- With more automation, the employees are spending too much on exceptions. There are not anticipated benefits. Begin to highlight the exceptions, as described in the chapter.

WHAT TO DO NEXT

1. Look around to find a process that was either implemented or improved in the past two years. Observe what has happened using the production process scorecard.
2. As you are doing this, try to detect if people are concerned about the process deteriorating.

PART V

Deal with Specific Issues

CHAPTER 13

Management Issues

INTRODUCTION

Part V addresses a series of common issues in different areas that you are likely to encounter as you consider and do process improvement. Viewed in a positive light, these are really general lessons learned. They represent a subset of the more than 200 issues that we have compiled in doing process improvement over the past 25 years. The ones selected were based on severity of impact, frequency of occurrence, and the ability to prevent the issue from occurring.

For each issue, four subjects are addressed:

- *Impact.* Here some of the major effects of the issue are identified.
- *Prevention.* This consists of some tips to prevent the issue from occurring.
- *Detection.* These are guidelines to help you detect when the issue is about to surface.
- *Action.* These are specific steps you can take if the issue appears.

Note that due to space limitations and the fact that the characteristics of each issue are unique in a specific situation, the comments are general.

It would be nice to say that there are no issues or that every process improvement effort is unique so that each issue is unique. But it is just not true. The same issues keep recurring due to the nature of people and organizations.

There are several ways to use the materials in this part of the book. First, you can refer to the materials for specific issues. Second, you can create a checklist of the issues to make management aware of what problems might surface when doing process improvement. Sometimes, if you warn someone of a problem, it has less likelihood of occurring. We have found that this leads to a much more realistic approach in carrying out change.

ISSUES

Management Changes Direction Often

Direction here refers not only to process improvement, but also to what management is devoting its time to. Change of direction tends to affect process improvement more due to the wide-ranging nature of process improvement.

- *Impact:* If management changes overall direction, it can divert resources from process improvement into other activities—thus, starving process improvement of resources. If process improvement was portrayed from a top-down perspective all along, then the impact is more severe, because there is less support at lower levels in the organization.
- *Prevention:* You have to assume that managers will change their attention and direction. Thus, we have learned and followed several basic guidelines. First, once process improvement starts, get grassroots support so that you are not overly dependent on management. This helps make process improvement self-sustaining. Second, do not draw attention to process improvement. Try to make it boring and dull. This gets you less management attention, so you are less impacted by direction change.
- *Detection:* You can detect management changes in direction through informal contacts. Keep several high-level managers informed about the effort. See what their level of interest is. Try to find out where managers are spending their time. Where they spend their time in new ideas or activities is a tip-off that a change of direction is coming.
- *Action:* If management does change direction, then you should adopt several actions. First, you continue the work. Second, you indicate that this is expected given that process improvement is not just a few Quick Wins. Third, approach management and volunteer to help the new effort by minimizing resources.

Lack of Management Will to Make Decisions

Process improvement often generates changes in policies and procedures. In more radical reengineering, there is organizational change. As you have seen, we deliberately do not adopt reengineering. In that way, we avoid organizational change. Organizational change can come later as a separate activity. Procedural changes can often be handled at lower levels. However, it is in the area of policy change where management may not be willing to make decisions.

- *Impact:* Failure to make decisions means that the existing policies remain in force. These may conflict with the new transactions. This causes difficulty and even a crisis in that employees may not know what to do about the work.
- *Prevention:* At the start of process improvement point out to management the potential policy changes that may be suggested. Propose several of these changes to see how management reacts. Indicate the effect of inconsistent policies and procedures. When you present the new policies later for approval, then you should focus not on the benefits of the new policy (which can be fuzzy), but on the negative impact of continuing the old policy. Focusing on the negative often brings about faster decisions.
- *Detection:* If you present a change and the need for decision and you do not get a fast response, you can assume that there is an issue or problem. Typically, this will be a political problem or issue.
- *Action:* If you find that management is not making a decision, do not confront this head on. You will lose. Instead, go informally to several key managers and indicate the impact of delays in the decision. Try to find alternative wording to address political concerns.

Turnover of Upper Management Is Substantial

A number of process improvement programs, such as Six Sigma and other efforts, fail because they depend on one key manager for support. When the manager leaves, support evaporates. All of the enemies of the departed manager sharpen their knives and take out their aggressions on the manager's project ideas. This has been going on since the time of Ramses I in Egypt.

- *Impact:* Process improvement is at risk with the departure of the champion. Progress will slow. Decisions will not be made. Work will flounder.
- *Prevention:* Do not depend on one manager for support. Even if one manager brings in process improvement, it must be embraced up and down the organization. After all, the managers are not doing the process work; the employees are.
- *Detection:* You can sometimes detect that someone is leaving if the person is spending time away from his or her normal duties. The person may be more distant and less involved.
- *Action:* If you find that a key manager is potentially going to leave, our approach is to go directly to the manager and ask him or her to help get other managers involved so that there is wider support. This, of course, should have been done at the outset.

The Managers Who Support Process Improvement Are Not Politically Strong

There is often a better understanding of business processes the further down in the organization you go. Managers who support process change tend to be interested in the work and often not as interested in politics. They are more hands on.

- *Impact:* People see the need for process improvement, but you can only do so much without upper management support. You will need management help when you face a political issue that moves up the organization. Managers who are more powerful will then not make changes.
- *Prevention:* Let's assume that there is a politically powerful manager who has no interest in process improvement. The manager may just want to further his or her own career, for example. In such cases, you appeal to the manager's self-interest and problem-solving interest. Don't treat process improvement as process improvement; treat the issue as an issue.
- *Detection:* Observe how people react when you go around to introduce the process improvement effort. Try to detect what people's self-interest is.
- *Action:* One effective method for dealing with a manager who has no interest in process improvement is to give attention to how the person's own domain or department might be impacted. This might get you more support. Keep a low-profile on the process improvement effort so that you avoid confrontation.

The Company Is a Startup, So It Lacks Resources and Experience

Go back to ancient times—1998 and 1999. Many startup companies were formed by people with good ideas, but a lack of process experience or knowledge. Implementing standardized business processes is difficult, because you cannot get management's attention. Managers don't see the importance of processes.

- *Impact:* Lack of decent business processes as much as flawed business models were the cause of the failure of many dot.com firms. When looking at failures, you repeatedly find a pattern of good ideas that were poorly implemented. Implementation here means putting in decent business processes.
- *Prevention:* For a startup you have to define specific, key business processes. Then you define the minimum, simplified transactions for each process.
- *Detection:* You can detect problems easily by following through sample transactions as a customer for a Web-based firm.

• *Action:* Some of the startups that survived brought in people who had process and operating experience. In other cases, the startup was folded back into the original company.

Management Wants Rapid Results

There is often a lack of patience among managers. They can understand that IT work takes time. But they often cannot understand why you just can't change a process overnight. Consultants who pitch reengineering and other methods often have to promise large-scale benefits to get approval to do work. This, in turn, raises expectations.

• *Impact:* High expectations and the drive for fast results translate into pressure for Quick Wins. One Quick Win is just to terminate X% of the staff based on seniority, supervisory, or manager opinion. In such cases, processes tend to deteriorate and not improve.

• *Prevention:* How do you prevent the rise of expectations and the impact of pressure for rapid results? One suggestion from experience is to surface many detailed issues to management. This will indicate that carrying out change does not come quickly. Next, take a sample change and develop a schedule for changing this one item. Go over it with management. This will give them a more realistic view of the world.

• *Detection:* One sign that pressure is coming is when managers seek to become suddenly more involved in the process improvement effort. This is a sign that they think that if they can become more involved, then they can speed up the work. They may want to assign more people to the work. That is like saying that if you want a baby in one month, you should get nine women pregnant at the same time. It obviously does not work.

• *Action:* Under pressure for fast results, you can focus management attention on issues related to Quick Wins. This will help the process improvement work and keep management busy with improvement issues. It will also buy some time for working on longer-term improvements.

Management Is under the Influence of Outside Investors, Consultants, or Other People

This is often the case in firms that started some new initiative, such as Six Sigma or e-business. Investors became excited with the change. The stock price rose. When results were not immediate, the pressure came on to get results fast.

- *Impact:* Outside pressure can be damaging to any internal effort. The outside people do not understand the internal processes. Management then puts on pressure for results.
- *Prevention:* It is clear that managers have to be careful about what initiatives they are launching since these now are much more visible to the outside world. Process improvement should be treated to the outside world as an ongoing activity. Quick Win results can help give credibility to the process improvement effort, if needed.
- *Detection:* You can detect this issue by reading about what investors and advisors are saying in reviews and on the Web.
- *Action:* There is no simple response here except to focus on Quick Wins. However, success will likely just spur more pressure to get more Quick Wins.

Management Wants to Select Processes That Are Not the Right Ones

This occurs when a high-level manager came from a specific area of the company and so has a lot of experience in one area. He or she may then want to devote attention to this area in terms of processes.

- *Impact:* If management decides what processes are to be improved, then there is likely to be fallout into other processes. These other processes may become dysfunctional or deteriorate. Benefits in the management-selected processes are offset by the deterioration in other processes.
- *Prevention:* At the start, you have to indicate how wide-ranging the choice of processes will be. You also need to disseminate examples and information on problems that arise due to selecting the wrong processes. It is also important to show how processes must be considered in groups.
- *Detection:* You can detect this when managers show too much interest in process improvement details. They may ask questions about the same area again and again.
- *Action:* Try to capitalize on the managers' interest in process improvement by showing them information from other processes. Use the process scorecards that were developed earlier to compare processes. Also, start bringing up issues related to Quick Win opportunities. This will, hopefully, channel their interest into positive areas.

The Company is Profitable and People Do Not See the Reason for Change

Things are going okay. Everything seems fine. Why change? This situation occurs frequently and represents inertia at the highest management levels. It characterizes many firms at the end of the 1990s.

- *Impact:* Issues are not analyzed as to their source. Symptoms are treated. There is no real desire to rock the boat with change. The fear is that if things change, then the financial picture or operational situation could worsen.
- *Prevention:* It is the same as with a child or pet. If things are quiet, then you can assume that there is a problem. The approach is to keep surfacing issues and to get attention for problems and their sources. You are not crying wolf by doing this, but you are reasonably pointing out potential problem areas.
- *Detection:* Look at what active projects and initiatives are going on. If this list is short, then that spells trouble.
- *Action:* In the face of complacency, you might begin to work on the process improvement effort at lower levels in the company hierarchy. People at lower levels of the company see problems every day so that they are more aware of what is needed.

Management Wants to Be Heavily Involved in Process Improvement

This can drive anyone crazy. Micromanagement is one of the curses that has been going on for hundreds of years. In several Six Sigma efforts, a high-level manager has gotten involved in transaction details. The results can be terrible.

- *Impact:* When a high-level manager gets involved in detail, the lower-level employees back off. They don't question the upper-level manager because they want to keep their jobs. Unrealistic changes are made in processes. The process deteriorates.
- *Prevention:* The best approach we have found for this issue is twofold. First, involve managers in detailed issues. This will get boring and they will tend to lose interest. Next, use the executive committee discussed in Chapter 2 to get them away from the detail. We do this as a matter of course in many process improvement efforts.
- *Detection:* When a manager starts attending process improvement meetings or starts to get too involved, you know that trouble is coming. Track the manager's involvement and interest from the start.
- *Action:* Follow the same steps under prevention.

There Is a Lack of a Company Vision

You have to wonder why a vision or mission statement is important. To many, it appears irrelevant since it does not relate to the work at hand. You saw in Chapter 2 how the vision and mission related to business objectives. If you lack a vision, then you also are going to have problems with business objectives. This cascades down to the work and processes.

- *Impact:* Everyone is focusing on short-term tasks. Get the work out. There is no long-term view. This will often mean that management will center their attention on Quick Wins and not on long-term improvement.
- *Prevention:* Formulate business objectives even if there is no vision or mission. Also, in the process plan presented in Chapter 6, make sure to give attention to the objectives of the process. These actions will help the light shine on the longer term.
- *Detection:* Not having any vision or mission is easy to detect. Having an out-of-date vision or one that no one is aware of is about the same. Look around and see where the vision or mission appears and how it is referred to.
- *Action:* Follow the steps under prevention.

Management Wants to Stop Improvement after Quick Wins Are Achieved

If the Quick Wins are substantial, there is a natural tendency among some managers to redirect the resources into other areas. Also, they may feel people are burned out and need a rest. However, the momentum for change is really underway.

- *Impact:* The tragedy here is the lost opportunity when change is stopped. Morale drops because employees feel that they were so close to achieving major change. They also realize that some of the Quick Wins need to change later.
- *Prevention:* Keep showing how Quick Wins lead to overall change. Process change in the longer term should always be linked to short-term efforts.
- *Detection:* You can detect this if you see that managers have become impatient with the results and are asking when the project will be completed. They may want to reassign some key resources as well.
- *Action:* Quick Wins must be presented as short-term steps to longer-term goals. It is important to show through the implementation strategy used in Chapter 8 how things fit together.

CHAPTER 14

Business and Organizational Issues

INTRODUCTION

From general management issues, this chapter considers potential problems at the business unit and organizational level. From experience, you are more likely to encounter these problems earlier in your work.

ISSUES

Business Departments and Managers Do Not See the Need for Change

The work has been going on the same way for many years. There is an established power structure. They see threats to the stability of the work.

- *Impact:* Resistance is often not stated or open, since upper management supports the change. A typical approach is that managers will raise issues and problems more subtlely, which delays change.
- *Prevention:* To prevent this from happening, you do not want to have upper management put pressure on the managers. A better approach is to gather grassroots support for change through the employees and lower-level supervisors.
- *Detection:* Look for signs of questions being raised. Another sign is for shadow systems, workarounds, and exceptions to be raised.

• *Action:* The early work of process improvement at the detailed level should concentrate on reviewing the current processes and uncovering problems in the work. As this is going on, you can also start getting the managers to acknowledge these problems.

Business Staff Lack Knowledge of The Processes

Many assume that everyone knows about why, what, and how they do their work. This is often not the case. Many organizations fail to train new employees who have had similar experience before. Over time, people may be governed more by habit.

• *Impact:* When people lack knowledge of why and how they do their work, it can be both a blessing and curse. Lack of knowledge makes collecting information more difficult. However, it can be an advantage because there may be less resistance to change with the new processes.

• *Prevention:* Better training can be a start. Another step is to measure and track processes more closely.

• *Detection:* This problem can be tracked down by visiting and meeting with employees. You can ask how they are working.

• *Action:* There is no simple action here. You do not want to retrain people in the old processes. However, you can gather lessons learned and implement other Quick Wins.

The Culture in the Company or Country Does Not Lend Itself to Innovation

In countries where there are low labor costs, there is sometimes a tendency to downplay improvements since the labor savings are not significant. In other countries it may be very difficult to terminate employees—another factor reducing change. There also may not a culture of change and innovation. This is not far-fetched at all. Many civilizations survived for a long time without any major change.

• *Impact:* People not only question the need for change as stated earlier, but they may have no experience in carrying out change. Even after change is implemented, there will be a tendency to revert to the original procedures.

• *Prevention:* This resistance to innovation is impossible to really prevent since it is not controllable. However, the company can make the effort to install new procedures when it first sets up the operation.

• *Detection:* This issue can be detected formally or informally. You can ask managers what changes have been made in the last two years. Informally, you can visit where the work is done and observe the work in progress and then ask about change.

• *Action:* You really have to go back to the first step—having people acknowledge the need for change. If people minimize the benefits, point to nonfinancial benefits related to productivity and morale.

The Best People To Be Involved in Process Improvement Are Already Overcommitted

This happens frequently. A department may depend on a key person to do some of the work. Or a key person may be assigned to another project.

• *Impact:* Having a critical person unavailable at all is a real issue. Fortunately, this is a rare event. The most frequent problem is that the person's availability is very limited. The impact is that you could miss some information, causing the work to be delayed.

• *Prevention:* Plan ahead for what you need. Minimize the time requirements for any one person. Try to avoid dealing with exceptions, workarounds, and shadow systems to the extent possible.

• *Detection:* You can detect this early when you are assembling the members of the action teams. You can also ask, "Who is busy on other work assignments?"

• *Action:* Try to use other people and build up a list of questions or issues for a critical person. Then show the person the list. Try to get the critical person interested in the work.

There Is a Resistance to Change Policies

It is one thing to change procedures. It is quite another to change policies. Some policy changes can have a major impact on labor requirements. Some changes have a direct financial impact. Many dread a policy change because they fear being blamed for any negative impact.

• *Impact:* The resistance to a change of policies tends to be more vocal and outspoken as opposed to the response to a procedural change. Even if you are enforcing an existing policy, there could be problems because some people may have interpreted the policy differently. The impact can be that the work is undermined.

• *Prevention:* Make sure that you are aware of what policies are actually being used in the organization to do the work. These may not be the stated policies or what management thinks is going on. Next, you want to ensure that policies are part of the analysis of the current process and work procedures.

- *Detection:* You can detect what policies are being used by having people demonstrate how they do their work through sample transactions.
- *Action:* Keep in mind that even if people support the new procedures and policies, it is not a sign of lasting support. You must continue to revisit departments after changes have been made.

There Is High Employee Turnover

In some situations there is high turnover due to wage levels, social factors, political problems, or other factors. High turnover may be due to the unpleasantness of the work or the working conditions. In one bus agency, new employees were placed in a customer information center. Due to a lack of automation and poor working conditions, everyone kept trying to transfer to better jobs in the agency. There was a 45% turnover every month.

- *Impact:* High turnover means that the process work is not consistent or stable. Things are falling apart. Managers may have given up on training employees due to the high turnover. High turnover, however, is not all bad. It can give you opportunities to make changes since there will be less resistance to change.
- *Prevention:* You can take steps to reduce turnover through wages, working conditions, and improved processes.
- *Detection:* High turnover is a known factor among management. However, on the surface it may be that the turnover is not high. In reality, critical people may be leaving in small numbers. So you must determine who is leaving.
- *Action:* Implementing improved processes and Quick Wins should help with reducing turnover. These changes can also bring more stability to the work itself. Since it received more management attention, people tend to pay more attention.

There Is Resistance to Achieving the Savings from Process Improvement

Here you implemented improved processes that increased productivity. Yet people have not been reassigned. There appear to be no significant benefits. This occurs more frequently than you think. Managers may be thinking that someone else will be responsible for getting the benefits.

- *Impact:* The first direct impact is that further process improvement efforts are questioned. Another impact may be to blame the process improvement effort for not getting the benefits. Either way, it is more difficult to move ahead. The situation also arises because managers have been promised major improvements and benefits as a result of many methods. These improvements have not materialized, so managers do not take the issue of benefits seriously.

- *Prevention:* Raise the issue of benefits and how they will be realized at the start of the improvement effort. Propose a situation in which "there is a savings of four people. What do we do with these people?" Use the early Quick Wins to prevent the problem later.
- *Detection:* Find out what happened when there were previous improvements. How were people reassigned? Ask the human resources manager or staff what happens in reassignment and how much reassignment has occurred.
- *Action:* If you find during the Quick Wins that no one is taking advantage of the benefits, raise the red flag and indicate that there is a problem.

There Are Many Shadow Systems in Departments

Recall that a shadow system is a set of procedures, internal systems, or other approaches to work that was created informally in the department. The shadow system has now become part of the normal means of doing business.

- *Impact:* Shadow systems tend to inhibit change, since they often must be adjusted when the processes change. However, it is likely that the people who created the shadow systems are long gone. Thus, the department is unwilling to give them up. This sparks more resistance to change.
- *Prevention:* The first step is to identify these shadow systems and get them out in the open. Second, you need to fix the scope of process improvement so that you can avoid dealing with as many of these shadow systems as possible. Otherwise, they can bog you down in the work. Third, include the shadow systems that you cannot avoid with the work in process improvement.
- *Detection:* You can detect shadow systems by observing how the work is done over time. Look for the use of small PC-based systems. Also, see what procedures people are following.
- *Action:* The best actions are those mentioned under prevention.

Departments Do Not Get Along with Each Other

There may be a history of bad feelings among departments. It could be that there were problems in the past and the departments blamed each other. Perhaps some managers in the different departments do not get along. The point is that the situation exists and no one has addressed it. One reason is that upper management may think that the behavior is natural and that it has little impact.

- *Impact:* When departments that have to share work fail to get along well, employees in the departments tend to cover their behinds by creating a paper trail for all transactions. One department might record manually when each transaction leaves its department for the next department. This leads to increased inefficiency and reduced productivity.
- *Prevention:* To prevent this from happening when you implement processes, you should try to minimize department boundaries and focus on the work itself.
- *Detection:* You can detect this problem by observing transaction work in the departments. Follow transactions and work between departments.
- *Action:* If you find this situation, then you first must measure the extent and impact of the rework, additional effort, and increased labor. Then you can demonstrate the problem to management. Try to fix some of these problems through Quick Wins.

Departments Have a Great Deal of Autonomy

This is neither good nor bad. Often, autonomy waves increase and decrease depending on what management theories are in vogue. When standardization and control are in vogue, there is less autonomy. When there is a desire for more accountability and for initiative, then there is more autonomy.

- *Impact:* Suppose that you want to make changes in a department regarding several processes. There are significant company benefits, but not many for the department itself. All that employees within the department see is that they have to do a lot of work and change and that they will be blamed for problems. The impact is to raise the level of resistance to change.
- *Prevention:* This can be headed off by defining how much autonomy departments have today. Raise this as an issue early in the process. When a change comes up, make sure you identify what is in the department's self-interest.
- *Detection:* You can detect potential problems when you surface the Quick Wins. Try to find one that will impact a department and see how members of the department react. That will give you a good sign of what you face later.
- *Action:* When you encounter resistance by a department, you have several alternative responses. You want to go to upper management only as a last resort. Instead, you want to get the lower-level employees in the department on your side.

Each Location Where the Work Is Performed Does It Differently

Every major area in a country has a unique culture. There may be language differences as well as cultural differences. These increase as you move up to the international level. Thus, it is not surprising that international firms run into situations where different procedures and policies are used in various countries to do the same work. This is less common in industries that rely on physical labor, such as manufacturing, and more common in the banking and insurance industries. However, it is a fact of life and something that has to be addressed in an international process improvement effort.

- *Impact:* The impact of doing the work differently may be positive in that people have found ways to adapt. Therefore, when you have to implement improvement and change in a specific region or internationally, you want to identify and consider local factors.
- *Prevention:* There is really nothing to prevent here. There are bound to be differences. You must identify what differences and adaptation are needed and why, however.
- *Detection:* The easiest way to identify differences is either to observe the work or to have lower-level workers in different locations work together to analyze their respective procedures and processes. This interaction will surface the differences.
- *Action:* Many issues can hide behind cultural differences, so you must be careful to have exceptions and differences not only identified, but explained at the start of the work process.

The Organizational Structure Does Not Fit Well with the New Processes

This lack of fit is not unexpected. After all, the organization supports and provides resources for the business processes. When the processes change, there is bound to be some impact on the people and organization doing the work. Previously, many people tried to predict the impacts on the organization and to change the organization in advance. This did not work very often because there were unpredictable impacts. In general, it is better to carry out organizational change after you have completed and measured the new processes. In that way, you will know what organizational issues have to be faced.

- *Impact:* If the organization is not functioning well with the new processes, then there may be lost productivity and redundant effort. There may be an emphasis on the wrong parts of the process and work situation.

- *Prevention:* This cannot be prevented. It is bound to happen to some extent and is a sign that the new process has taken root.
- *Detection:* You can detect this issue when you measure the new process and observe how people do the work differently. Employees at lower levels will tend to surface issues.
- *Action:* You first want to have employees surface issues that relate to the new process. See if these concerns can be traced back to the organization. Then you can raise these and other issues with management, indicating that some organizational adjustment is necessary.

CHAPTER 15

Work Issues

INTRODUCTION

Work issues pertain to the more internal aspects of process improvement. While management and other issues tend to be considered more political, internal work issues can be more damaging if not addressed because they are linked to the people performing the process or involved in the process improvement.

ISSUES

There Is a Lack of Formal Procedures

At first, one wonders how this can be. If the work is organized, it must be according to some specific procedures, right? Wrong. At one time there may have been procedures, but they may no longer be followed. There may be no reinforcement or measurement. People tend to revert to informal procedures and to then run to the senior employees who have been there a long time and know the exceptions.

- *Impact:* This is both bad news and good news. The situation offers an opportunity to implement improvements. Also, it shows that the employees are sufficiently intelligent that they do not need the formal procedures. On the negative side, there tend to be more exceptions to deal with, making the process improvement effort larger than originally planned.

• *Prevention:* How do you prevent falling into a trap? When you first review processes, look for the existence of standardized procedures. Don't just ask for documents. Look at what people are doing. Ask them how they were trained and what rules they follow. You may then want to move on to another process.

• *Detection:* Detection of the lack of formal procedures follows from the preceding paragraph. Even if there are formal procedures, who is to say that they are still valid or being followed? Here is a test. See if you can do the work yourself by following the procedures. If you can, then you have validated the procedures.

• *Action:* One of your first actions will be to analyze the current procedures. This gives you the opportunity to create new, temporary formal procedures for major transactions and common work assignments. This can be a useful Quick Win. It is relatively easy to do since you are already analyzing the process.

Training of Employees Is Not Organized

People make a big deal out of training in systems and technology-based work. This is because this type of work is viewed as complex. There is an assumption that people are not familiar with computer systems. When you move to the business side of the work, the situation changes. The people who were hired may have had experience in similar areas, so there is a presumption that they already know how to do the work. Employees come and go. It is difficult, when you are trying to get work done, to organize training for one or a few people. For these reasons, the training of employees is often not organized well and may not even be done. On-the-job, hands-on training is substituted.

• *Impact:* Without training, there tends to be a lack of consistency in the procedures and policies being followed. How the work is being performed may be impacted more by the style of the individual supervisor. Another impact is a loss of efficiency as employees go to other employees and supervisors more often with questions. More exceptions creep into the work. The impact on the process improvement effort can be significant in that you may not get a standard or consistent description of how to do specific tasks. There may be major differences among supervisors as well.

• *Prevention:* Beyond the initial training, the best approach is to have sessions for the employees during which time they can share experiences and methods on how they handle the work procedures. This will facilitate knowledge transfer as well as give the employees more confidence in their work. It is a good Quick Win to prevent problems later.

• *Detection:* You can detect the training issue by observing the work on different shifts or under different supervisors. Have the supervisors train you in how to do the work. This will bring problems to the surface.

• *Action:* When you encounter this training issue, the first thing to do is to get the supervisors and employees to recognize that it is a problem. Often they will get defensive because they may feel that they will be blamed for the lack of training. Sympathize with them by indicating that no one can expect them to do all of the work, while at the same time doing the training. After this, you can create some short group sessions to share lessons learned.

THERE IS A LACK OF ACCOUNTABILITY FOR THE WORK

What does this mean? People are doing the work so aren't they accountable? Not necessarily. Work can be organized in a confusing way so that many different hands touch transactions and pieces of work. In such situations, there is a lack of both control and accountability. Of course, the employees may like it that way, since they can only be blamed for problems as a group. This situation is not only common in developing nations, but also in some standard business processes in Europe and the United States.

• *Impact:* Without accountability, there is also a lack of individual rewards and morale boosting of individual employees. The group may be praised, but the manager cannot single out one person. The impact is that productivity may slide to some minimally acceptable level. This situation also makes the improvement effort more complex and difficult.

• *Prevention:* To prevent this issue, you would have to restructure the work situation. Here you would cite the accountability and control problems as issues in the business process.

• *Detection:* By observing the work process, you can detect the handoffs of transactions between different people. Also, watch to see how errors are detected and fixed and, more generally, how rework is performed and organized. The fixing and rework can reveal a great deal about accountability. Another sign of problems is the repetition of the same problems without the underlying problem being fixed.

• *Action:* Once you have uncovered accountability problems, you have to be sensitive and careful as to what you do. If you try to step in and make changes to improve controls, then you are likely to be met with defensiveness. Don't make changes here in an effort to achieve a Quick Win. Instead, raise the level of awareness by showing the impact of the issue on the individual employees, in that they cannot easily receive recognition. This will help them see that there is a problem. Then you can involve them in defining the new processes.

Methods for Improving the Work Have Been Tried before, but the Results Are Not Evident

In several companies we have been involved with, there had been previous efforts at reengineering and quality management. Managers at higher levels even have told us that those processes were already fixed, so we should expect little additional improvement. Yet when we went out to observe and study the work environment, we found as many or more problems as there would have been if there were no previous effort. What happened? One thing that occurs is that there is no follow-up or measurement of the work after the project. So things gradually revert to the earlier state. This brings us to the fundamental truth expressed in the early chapters, which is that processes tend to deteriorate over time. Management often assumes that the processes are stable if fixed. But this is not true unless they are highly automated—like e-business transactions.

- *Impact:* There are a number of potential impacts. First, the situation may actually be worse because now a mix of newer and older methods are being followed. This can reduce productivity and raise confusion. Next, employees may feel that they spent a lot of time on the previous effort, which went nowhere. Now they are reluctant to commit to improvements since they think it will be another failed effort.
- *Prevention:* In retrospect, the key notion is that you must regularly measure and assess current business processes—even if everything seems to be fine. If people are aware of recurring measurement, you can keep the changes active and in place.
- *Detection:* While you can observe the work, this method of detection is probably too slow. Instead, go to the more senior employees and find out what was tried out before and what happened. Indicate that you don't want to repeat the errors of the past.
- *Action:* If you find this problem, you may have a major opportunity for Quick Wins as well as improvement. You might, for example, temporarily restore the old process as it is at least consistent. You can also make minor changes to keep some of the good parts of the previous work process. It is important that you address the concern that your process improvement effort will be just another failure. Another action is to get the employees together and gather lessons learned and experience on why the previous effort did not work out right.

Much Has Been Invested in Technology, But the Results Are Mixed

Many companies tend to throw money at business processes by overinvesting in technology. The feeling is that if they automate the work, any problems will be fixed. The failure of many e-business efforts shows that this is a bad idea. Technology and systems are only tools. They are not methods. It is like giving a child a bicycle—just because a child has a bicycle does not ensure that she or he can ride. The child needs to be taught the method of riding the bicycle.

- *Impact:* There can be many bad feelings here. Managers feels that they did a lot by investing in the technology. They feel let down and may be reluctant to make any further investment—even if it is needed for process improvement. Now consider the employee's point of view. The employees were given PCs or systems and not really told how they were supposed to use them in performing their work. This happened in the 1980s with earlier stand-alone PCs. Productivity declined as people played games or worked around the systems or technology. Overall, there can be a great deal of unhappiness.
- *Prevention:* Prevention here rests on the technology being only part of the overall solution for improvement rather than the improvement itself. That is why to prevent this problem, technology should follow the methods and support them—not lead the method. Otherwise, the method will likely be warped around the system and the technology.
- *Detection:* You can detect this problem by asking employees how they are using the technology. They can indicate how the technology is used through sample transactions. Take a similar approach when looking at how other kinds of technology are being applied in the workplace.
- *Action:* A potential Quick Win is to figure out how the technology can be employed better or, in some cases, why the company might stop using it altogether. You also want to bring up the use of the technology when you assess the current process.

The IT Group is Overwhelmed with Projects and Other Work and Cannot Support Process Improvement

The information technology group is like any other organization. IT has its own priorities and work and cannot drop everything to address needs due to process improvement. The situation typically is even more complex than this. The specific changes that you require to support process improvement are obviously linked to specific systems. Each production system is maintained and supported by specific programmers. Doing the maintenance work requires substantial experience with the system as well as specialized knowledge. Thus, no other programmers can really make changes in a realistic time frame. If these critical programmers are very busy, there is a problem.

- *Impact:* The lack of resources can delay the process improvement work. This may limit your effort to achieve Quick Wins until resources are freed up. However, you cannot expect to get lasting improvement in many cases without system changes or enhancements.

- *Prevention:* There is no reasonable way to prevent this situation since it is fact of life in most organizations. To prevent the impact, you must use the IT workload as a factor in determining which processes to improve.
- *Detection:* Detecting the problem is fairly easy. After you have identified some potential business processes, you can approach IT and see what the status is for the people working on the systems that are supporting these processes.
- *Action:* An early action is to begin to negotiate with IT management for resources and to set up a schedule for doing the work in the future. You should also keep IT informed about what is happening in the process improvement work to give the IT department more lead time.

There Are a Number of Projects Going on Already so That These are Receiving Attention

Many larger organizations have additional projects beyond process improvement that are demanding attention. Some of these projects were started before the process improvement effort began. Others began after the process improvement work started.

- *Impact:* The direct impact falls into two areas. First, there is competition for management and employee attention. What gets the time? The second impact is in the area of resource allocation. Process improvement already had to compete with regular work assignments for people's time. Now, in addition, process improvement must compete with additional projects.
- *Prevention:* This problem can be prevented only by addressing the various projects underway with management. At that time, you can bring up the resource and attention issues. Some work may be deferred.
- *Detection:* You can detect if there is going to be a problem in two ways. First, you can indirectly find out from employees what other demands there are for their time. Second, and more directly, you can talk to the project leaders of these other projects and work assignments.
- *Action:* The action you take must be proactive. As soon as you are aware of other projects and work, you should begin immediately to address the resource allocation effort in a proactive manner.

Processes Depend Heavily on Work by Suppliers, Contractors, and Consultants—in Part Due to Outsourcing

You may have selected processes that were partially or wholly outsourced. Alternatively, the processes you have selected for improvement rely or interface with processes that involve outsourcing vendors.

• *Impact:* This situation requires more coordination and planning. An outsourcing vendor has its own agenda and priorities—just like the IT department does. You could complete a great deal of the process improvement work, but be robbed of the benefits because you are waiting for the vendor to establish its end of important interfaces.

• *Prevention:* To head off problems, identify interfaces with other systems and processes. Include potential problems in the early analysis and selection of the processes for improvement.

• *Detection:* Having selected processes for which there are dependencies, you should be vigilant to detect any delays or situations where there are insufficient resources.

• *Action:* The proactive approach is to go directly to the vendors early in the project and negotiate for resources and for a schedule.

Resources Are Robbed from Process Improvement for Other Work

This relates to the issue of other projects and work. In this case, you were promised resources for the process improvement effort and were told it would be given priority. Later, new problems in other areas arose and management took resources away from process improvement. This is not uncommon given that some process improvement efforts take considerable time to complete.

• *Impact:* The obvious impact is that with the missing resources, the improvement work slows down. There may also be a decline in morale as the employees and team members sense the shifting priorities. After all, they may be thinking, "Why should I spend time on this if management no longer thinks it is a priority?" This is a natural response and one that you should assume will occur.

• *Prevention:* While you cannot totally prevent resources from being taken, you can take a number of steps to mitigate damage to the effort. First, you can probably predict which people will be in demand. Then you can schedule their time toward the front end of the process improvement work. Second, you can try to avoid involving individuals who are in high demand.

• *Detection:* You can detect when people are feeling pressure to work on other things by visiting their offices and seeing how they are doing. If someone begins to take longer to do the work on the project, then this is another sign of problems.

• *Action:* If someone is taken from the improvement effort, it is usually not a good idea to try and get the person back. This will raise management concerns that you are overly dependent on one person. Instead, work informally with the person to identify an alternate person and to turn over some of the knowledge and work to that person.

Decisions Are Often Made on an Ad Hoc Basis without Analysis

In process improvement and other significant projects, there is often pressure to get results. This means that when people are faced with issues, they tend to make decisions faster. They want to get the ball rolling and have actions taken. There is also the fear of paralysis through analysis.

- *Impact:* Decisions made without analysis have several potential negative impacts on process improvement. First, the decision may later have to be reversed based on new information. This results in people questioning the credibility of management and the work. Second, decisions may be based on symptoms of a problem or only a limited view of an issue so that other related problems now surface. The problem and situation are made worse.
- *Prevention:* You can prevent this from happening by encouraging the discussion and analysis of issues and not putting pressure on management for resolution. In process improvement, it is often wisest to wait and see what happens.
- *Detection:* To detect ad hoc decision making, look at what happened as a result of the decisions. What actions were taken? Do they still make sense? In retrospect, if you had to do it over again, what would you do?
- *Action:* If you find that decisions are being made too rapidly, then you should slow down the pace. You can indicate that there is not immediate urgency to solve every issue. You can also group issues so that people are more reluctant to jump in and make decisions.

The Process Improvement Work Gets off to a Slow Start

Like many new initiatives, process improvement can be started with great fanfare by management. However, management may overlook the realities of the schedule. People may have to work on other items.

- *Impact:* This can take the steam and momentum out of the work and force a relaunch of the process improvement effort later—not a good sign. Another impact is that people feel that management is not aware of the business conditions and situation when they set the wrong date. There is another impact—management still counts this down time against the schedule for completing the change.
- *Prevention:* The key to preventing this problem is to better analyze and plan when to start the process improvement effort in conjunction with other projects and work. This means that resource planning must be part of the analysis at the start.

• *Detection:* You can detect if there are problems at the start by finding out that individuals are not available to do the work. You may find that you have to search for alternative players more often.

• *Action:* If this occurs, you have several options. You can delay the work—not a good sign. You can proceed in a minimal way and limp along. Again, this is not desirable. A third approach is to use the time for more analysis so that management does not count this as time toward the completion of process improvement.

Much of the Work Is Being Processed as Exceptions

You have seen the factors that give rise to this problem in departments. Exceptions mean that each transaction is given close scrutiny instead of being processed directly. More time and analysis are needed. Emphasis falls on accuracy rather than productivity.

• *Impact:* An exception-based activity is much more difficult to improve. You cannot automate or streamline all exceptions—there are just too many. The danger is that you can spend many hours trying to improve things only to find out that, overall, there are little or no benefits.

• *Prevention:* Be careful about which processes you select. Try to determine why there are exceptions. Is it due to the nature of the work, how it is organized, or policies?

• *Detection:* You can easily detect the volume of exceptions by observing the work. You can also be trained in the work activity so that you will be immediately aware of all of the exceptions that are involved.

• *Action:* If you are forced to deal with such a process, you first should attack the exceptions in general to see if many can be eliminated. Failing this, you next attempt to identify the most frequent exceptions and address those.

New Opportunities Surface That Are Not Part of the Original Scope

This is bound to happen as you start to turn over rocks in departments. The opportunities appear exciting and can yield benefits.

• *Impact:* These new opportunities can be very enticing. They seem to be such good Quick Wins that work stops on the main effort and is channeled here. The schedule slips, and while managers think the other benefits are nice, they are resentful because you did not stay focused on the core target.

• *Prevention:* You and the team must realize that if the early work in the improvement effort is successful, then these opportunities will appear. Expect them, and channel them into Quick Wins for later.

• *Detection:* You can detect if you are being diverted by these opportunities by looking at where you are spending your time. If you are spending more time on these new opportunities than on core activities, then you are in trouble.

• *Action:* This is why we have Quick Wins. You can put these new opportunities into categories of Quick Wins to be addressed later.

EMPLOYEES ARE ATTEMPTING TO GO BACK TO THE OLD PROCEDURES AFTER CHANGES HAVE BEEN IMPLEMENTED

You implement the new methods and even systems. Everything appears to work. You measure the improvement process and the benefits are there. Now you and others go on to other work. The employees are left to their own devices to face new situations and work. By instinct, they often revert to the old methods. The new process is undermined.

• *Impact:* When reversion occurs, the entire process improvement effort is called into question. Why do all of this work and incur this expense if the benefits are not lasting? There is a good chance that the process improvement work will be stopped.

• *Prevention:* You must assume that this will happen. You will need to work with supervisors and managers to define procedures for overseeing and tracking the work. This is almost as important as the new processes and work flow itself.

• *Detection:* You can detect that reversion has occurred by visiting the employees and observing how they handle exceptions and new types of work. This is the start of the reversion.

• *Action:* If you find this occurring, you must go back into the department and fix the situation by handling the exceptions and deviations. Then you can implement the prevention steps stated earlier.

CHAPTER 16

Political and Cultural Issues

INTRODUCTION

Many books on process improvement do not address the issues that you will face when you implement improvements. The few that do fail to deal with the real world of politics and culture.

ISSUES

Each Country in Which the Company Is Located Has a Different Culture

Most organizations often assume that processes must work basically the same way in any country. This was not true in Roman times. The Romans had to adopt their laws and method of government to the culture of each group of people they ruled. A faulty assumption is that if you manage something, you can define in detail how to do the work. This is true only if you stay in that country to direct it.

- *Impact:* Let's take an example. A company implements a new sales or customer service process across a region. The team leading the implementation is based at headquarters. When faced with cultural issues, team members often indicate, "Well, it works elsewhere doing the work the same way. It should work here the same way." The impact can be catastrophic. The process may actually be implemented the same way, yet sales are lost and customers go elsewhere. Alternatively, the process reverts to an older form, wasting the implementation effort.

- *Prevention:* To prevent this from happening, it is important to dry run the new process in each country to surface potential problems. As you uncover problems, you can then identify versions of the process that are individualistic. However, these will be consistent overall across the countries.
- *Detection:* You can detect this problem by visiting and comparing how work is done. Have people show you how they do their work. Probe for differences due to language, culture, religion, and so on. This will tip you off to what you will face later. Do the same if you are in the middle of implementing a new process.
- *Action:* If this problem arises, you should begin a new part of the work to identify cultural issues in the different countries. That is why when you implement a new process, you should try to go to the country with the culture that is most different from your own. This will reveal the worse case.

IT and Business Departments Are Hostile to Each Other

Over many years, information technology has provided support for the business units and management. IT has been put in the middle between management's desire for something and departments resisting the change based on not seeing the need for it. IT, in many cases, has been the reluctant agent for change. As one IT manager said, "We signed up to deliver systems; we did not gear up to deal with emotional, political, and other issues involved in change." Add to this the different orientation of IT and business units, and you can see why there is some hostility and mistrust. IT wants to have exact requirements to work with the technology; the business considers requirements to be flexible depending on the situation. IT works toward specific schedules, while the business department sees work as continuous.

- *Impact:* Problems between users and IT can get in the way of getting processes and transactions changed. Work can be delayed. Members of the IT department may be unwilling to devote resources to help the effort if they perceive that the users will not follow the new process, for example.
- *Prevention:* Get over the idea that any problems between IT and business units can be fixed quickly. It will take time, given how long the relationship has been going on. So you cannot prevent this problem. What do you do to mitigate it? Work at lower levels in both IT and the department. It is at lower levels, where there is daily coordination, that you will have the most success.
- *Detection:* You can detect this issue by talking separately with users and IT managers and staff. Ask them how requests for new services and support are handled. Ask for examples.

• *Action:* When the poor relationship begins to affect the work and progress, you should consider a two-pronged approach. At the lower levels, continue to work with individuals in both organizations to keep working ahead. At the higher level, you want to present the problem to managers. However, do not overemphasize the effect on the improvement effort—this will sound like you are whining. Instead, point to it as a general problem with many negative effects.

Upper Management Is Divided into Different Political Groups

Upper management often functions as a team. But when people get into the detail, they often reflect the bias, attitudes, and opinions of their own organizations. It is in these situations that politics may enter. You should assume that much of this is natural. There is a healthy tension between marketing and accounting.

• *Impact:* If you are not aware of this issue and do not deal with it, the improvement effort can be caught in the middle. For example, suppose that you have a transaction that spans two departments. If the department managers are not cooperative, then this will be reflected in the attitudes of their respective employees.

• *Prevention:* This is another situation that you cannot prevent, so you must adopt more subtle techniques to get around the problem. Wherever possible, work at lower levels of the departments where there tends to be less hostility. Do not hold joint meetings with the departments at the same time until you are almost certain that there will not be a problem. Otherwise, a simple meeting can retard progress for weeks.

• *Detection:* To detect this problem, review specific transactions to see what happens as you follow the work across and within departments. Ask what problems employees encounter when they receive the work from the preceding department. Then go back to the sending department and ask what complaints they hear from the receiving department.

• *Action:* Knowing that this situation will often exist, you should begin by assuming that it does exist. Look for it using the tips in detection. Then assess how severe it is. Start making people aware of it as a potential issue. Notice we said a *potential* issue. You do not want to raise it as an issue. You want to give managers an opportunity to think about it and, perhaps, address it on their own. If the problem arises, you can refer to what was said earlier.

Department Managers and Others See Change and Process Improvement as a Threat to Their Power

This occurs because managers and supervisors have become accustomed to a pattern of work. It occurred in the Renaissance period of history when the element of time was introduced. Before clocks appeared, people worked at a slower pace. When time was introduced, processes changed. This created resentment among many people. In some towns, clocks were destroyed as the "work of the devil."

- *Impact:* What is the impact of this problem on the improvement effort? Obviously, it may cause resistance. However, much of the resistance will not be in the open. It will be covert. Supervisors may raise many problems and issues and ask how the new process and transactions will handle them. Be on guard. Carried to an extreme, the department might revert to the old methods after the improvement team leaves.
- *Prevention:* How do you prevent this from happening? First, assume that this issue is present and raise it openly. Indicate that you understand people's feelings and that these are natural and part of human behavior. Let everyone know that you will be sensitive to these feelings and address their concerns as the work progresses.
- *Detection:* Detection comes through suggesting potential changes and seeing what reactions people have. This must be done before you implement change. In that way, it is politically soft and not as threatening. Propose major potential changes as just thoughts or ideas to test how strong their feelings are.
- *Action:* When you encounter resistance, indicate that you are not addressing organizational structure. Your focus is on how the work is performed. Try to make the issue nonpolitical by surfacing the politics in an indirect manner.

Suppliers Are Resistant to Being Involved in Process Work with the Company

This arises in e-business. Suppliers have their own agendas and priorities. You are trying to implement e-procurement, for example, and need their participation. If you try to be heavy-handed because the supplier depends on you for much of its business, your actions will backfire sooner or later. This negative approach has been going on for centuries as larger firms try to coerce their suppliers to do things in a certain way. It really fails unless there is some benefit to the suppliers for implementing the change.

- *Impact:* The impact is that the suppliers are slow to participate. The schedule for e-business slips. The entire effort is in jeopardy.
- *Prevention:* To prevent this problem, you must understand the concerns, desires, and self-interests of the suppliers. This requires spending time with them. You should do this at the beginning and not in the middle of the e-business effort.
- *Detection:* You can detect the problem in several ways. The supplier may cancel meetings regarding coordination. The supplier's employees may not perform

their agreed-upon tasks on time. The supplier's key people may not be available; instead, more junior people are assigned. These are all signs of problems.

• *Action:* Even with the best of intentions and effort, this problem will likely occur sometime during the e-business effort. The improvement effort just takes too long. People will be drawn off into urgent, short-term work. The action is to assume that this issue will arise and be watchful. When it occurs, meet with the supplier managers and review the problem and the schedule. Be proactive. Do not assume that this problem will fix itself. It will not. In fact, many times it will get worse as the supplier's employees feel that you have changed your priorities too because you are not following up.

Customer Needs Are Not Clear

Many people assume that they know what customers want, so they design processes and systems based on these assumptions. But they make many mistakes. They assume that customers want more information; they don't. They assume that customers will behave in a certain way; they don't.

• *Impact:* Without analysis and testing of customer needs, many new processes and e-business efforts will not be used. That is one reason for a substantial number of e-business failure—the company misjudged the customer.

• *Prevention:* To head off this problem, you need to look at customer needs from a very simple view. In some cases, we have had teenagers act as customers to determine basic needs. Work to define the basic needs that must be met in order to complete a transaction. You can always add more features and capabilities later.

• *Detection:* You can detect that there is a problem in the testing and evaluation of requirements for the new transactions. If the reviews and tests are performed by internal employees, then the results can be questioned.

• *Action:* If you see this problem arising, go back to the original requirements and push for a review to ensure that the work is on the right track. Press for a reality check.

There Is No Existing Culture to Draw upon Experience and Use Lessons Learned

This happens with new organizations and startups. It can be as damaging as resistantance to change. You are working with somewhat of a blank slate.

• *Impact:* The impact is that you must be proactive to define basic processes and transactions. There is not sufficient time to cover every single detail. Some problems will have to be worked out as the processes get underway. Now any new organization will hire people who have previous experience. These individuals bring with them good and bad baggage from their previous jobs. They have good practices and bad habits. You must work to ensure that the good practices come through and the bad habits do not. The impact is that the employees will implement the old processes—good and bad. Then, since there is not the history or past to draw upon, the bad will tend to survive.

- *Prevention:* Preventing the problem means that you must be heavily involved in setting up the new transactions and work. Here is what to do. For each piece of work, get the employees together in a group and ask how they did it in their previous jobs. Work to get consensus on how to approach the work. This will deliver several benefits. First, it will support good practices. Second, you are building a collaborative culture by having different employees share their experiences.
- *Detection:* You can detect problems when you observe how the work is being performed by different people at different times. You can detect inconsistencies and then probe for the reasons why the work is being done in one manner instead of another.
- *Action:* You are going to have to counteract bad habits from the past to keep them from intruding into the processes. This is not going to be addressed in the procedures, but in the practices for how the work is done. Encourage the sharing of lessons learned on a regular basis to keep the practices clean.

There is No Compelling Need to Improve Processes Now

The company may be successful. It is making money. There is low turnover. There appear to be no clouds on the horizon. Why rock the boat? That is a major reason why it is difficult to implement change in good times. It was true in ancient Rome and Egypt; it is still true today.

- *Impact:* There can be much resistance to change and improvement when times are good. People will say, "Why spend time on improvement when we could spend the equivalent time doing more work?" This is understandable. There will not only be resistance, there will also be more of a tendency to revert to the old, proven ways once you disappear.
- *Prevention:* If you are going to implement improvements in good times, there are some specific guidelines to help you out. First, identify potential clouds on the horizon. Point to the impact of not keeping up. Second, you can give examples from other industries or firms in the same industry in the past. Overall, you are trying to raise fears, to get people out of their complacency.
- *Detection:* You can detect this problem if the improvement effort is generated from the top down and meets almost immediate resistance from everyone else. You can also detect it by proposing changes at the start.

• *Action:* You should have people indicate what problems they face in performing their work. What would they like to see done better? Getting this information out on the table will help you encourage change.

There Is a Tradition of Political Infighting within the Company

In a number of organizations we have both worked in and consulted with, there was a long history of infighting. In several cases, top management encouraged this behavior as a way to get the best ideas out on the table. It is like putting two animals in a ring or arena and seeing which one wins.

• *Impact:* Excessive infighting can derail process improvement almost from the start. Every time an issue or opportunity is brought forth, people take opposing sides. There is no progress. Improvements are either delayed or do not occur.

• *Prevention:* While you cannot counter the tradition, you can take some positive steps. First, you can informally acknowledge that the problem exists and even ask for suggestions from managers on how to get around it. Infighting typically occurs at higher levels of management. Therefore, you should keep to the lower levels of the departments where people are too busy to fight.

• *Detection:* You can detect this problem by interviewing and talking informally with various managers about similar topics one on one. This approach will surface different opinions. You can then propose some specific ideas for change and ask how they would implement the change. This discussion can then lead naturally into implementation issues.

• *Action:* Work at lower levels where there is less infighting. Acknowledge with managers that it is natural for people to have different opinions on a subject. Keep any presentation or discussion to the details related to improvement. Try to keep other subjects, such as organization, out of it by stressing that these issues are not within the scope of the improvement effort.

Roles and Responsibilities in the Organization Are Not Clearly Defined

Often, this issue of unspecified roles will be present in smaller organizations or in branch or regional offices. When responsibilities are not clearly defined, there is a vacuum that must be addressed or at least acknowledged.

• *Impact:* The direct impact is that when you begin the improvement effort, people will tend to become confused and not want to be involved. This can slow down the improvement process.

• *Prevention:* You can prevent many problems here by focusing on detailed transactions. Take a sample transaction and simulate the roles of the people in a meeting by going through the analysis.

• *Detection:* Try to detect this problem early by following the guidelines under prevention. You can also detect the problem if you find that people are not performing their assigned and agreed-upon tasks.

• *Action:* When the problem surfaces, it is a good idea to pick a transaction and go through the steps noted under prevention.

An Individual or One Group Seeks to Take Over the Improvement Effort

You may have found that people were not interested in the effort at the beginning. This occurs for many reasons that have already been discussed. They are busy. They do not think anything will come of it. Then when they see results and momentum, many people will try to jump on the bandwagon to show that they are on the team. Also, some managers may try to take over the change effort.

• *Impact:* The direct impact is that you must now deal with individuals who really want to be involved. They want to be in on the decisions about the work. Sometimes, this can be positive. At other times, this can be negative and slow the improvement effort down as these individuals attempt to channel the improvement effort in different directions.

• *Prevention:* You can prevent this problem by raising it as an issue early on. You should emphasize the roles and responsibilities and the method we presented for coming up with the new processes and setting the implementation strategy.

• *Detection:* You can detect this problem if you suddenly observe managers becoming very interested in the work. This is a tip-off that problems lie ahead.

• *Action:* When this issue occurs, review the process and roles. Do not turn people off. They will become hostile and then other problems will arise. Instead, try to channel this enthusiasm toward implementation and potential implementation issues.

Some Employees Are Unionized, Making Change More Difficult

We have carried out many improvement efforts in bureaucratic and unionized structures. People may think it is impossible to deal with this issue. It is not. It is just different and challenging in its own way.

• *Impact:* In a highly organized and structured environment change is more difficult. You have to consider that it will take longer to get formal decisions and approval. The improvement effort slows down.

• *Prevention:* In such an organization, one approach that has worked for us is to keep things informal. When you make things formal, you tend to encounter more resistance. You can also work with individual union members and the shop steward to get them on your side for change. Remember that they would have to have improvements too.

• *Detection:* You can detect specific problems when managers indicate that they cannot decide or make decisions without more review. This is a sign of bureaucracy.

• *Action:* When you encounter this type of resistance, back away and move to informal channels. Try to surface a problem or issue early to see what the level of resistance will be. This will help you in planning.

There Is No Culture for Measurement or Justification of Change

In some organizations people just introduce changes without analysis. So when you come along with an organized approach, they wonder and question why this is necessary. After all, they might say, "We have always just made changes that made sense. Why don't we continue to do it the same way, since it works?"

• *Impact:* One potential impact is that there will be more pressure to implement change. If you join this bandwagon and dump your method, it will be harder and even impossible to recover later. Another impact is that the changes that are implemented will be inconsistent.

• *Prevention:* To prevent the problem, first present the method and give examples. You can also indicate that you understand how many would want to rush out and make changes. Point out the problems that have occurred in the past when this was done.

• *Detection:* If you see that people jump on the Quick Wins, then you know that the problem is present.

• *Action:* If someone wants to just implement a new method quickly, point out the ramifications of change and impacts. How will the company deal with the side effects of change, for example?

APPENDIX 1

The Magic Cross-Reference

The Magic Cross Reference is intended to save you time instead of using an index. Major headings appear in the first column. Detailed topics appear in the second column. The relevant page numbers appear in the third column.

Area	Topic	Pages
Implement improved processes	Organization	199–200
Implement improved processes	Policy changes	195
Implement improved processes	Prototyping and piloting	191–195
Implement improved processes	Systems related	195–197
Implement improved processes	Training and documentation	200–201
Implementation plan	Construction	164–165
Implementation plan	Methods	155–159
Implementation plan	Template	160–163
Implementation strategy	Improvement tables	143–144
Implementation strategy	Planning	140–143
Measurement and Maintenance	Benefits validation	216–218
Measurement and Maintenance	Measurements	214–216
Measurement and Maintenance	Organizational change	218–219
Measurement and Maintenance	Process coordination	210–212
Measurement and Maintenance	Process deterioration	219–220
Measurement and Maintenance	Process enhancement	212–214
New business processes	Alternative group of processes	66–67
New business processes	Process group score cards	70–71
New business processes	Process group selection	69–70
New business processes	Process relationships	64–66
Planning	Assess business objectives and issues	37–38
Planning	Business related tables	39
Planning	Evaluate mission and vision	36–37
Planning	Major deliverable items	33
Planning	Organization score card	39–40
Planning	Planning steps for improvement	29–30
Planning	Process improvement organization	32–33

APPENDIX 2

Improvement Tables

This appendix summarizes the improvement tables in terms of categories and purposes. This appendix has proven useful in a number of improvement efforts by providing a road map as to what tables might be of use at the start of the effort. We begin with lists of factors that are then employed to produce the tables. Remember that you do not have time to develop all of these improvement tables. Instead, you will be selecting the ones that are valuable for gaining support and understanding of your process improvement effort.

FACTORS

Business-related factors

- Business mission
- Business vision
- Business objectives
- Critical business processes
- Business issues
- Organization

Systems and technology factors

- IT objectives
- IT issues
- IT architecture

- Potential technologies and systems

Process group–related factors

- Process group
- Individual processes

Individual transactions

- Process steps

Current transactions
Long-term transactions
Improvement implementation strategy

- Quick Wins in a phase
- Future process changes

IMPROVEMENT TABLES

Type	Improvement tables	Comments
Business	Business vision vs. business mission	Consistency
	Business vision vs. business objectives	Follow through on the vision
	Business objectives vs. business issues	Impact of issues on objectives
	Business mission vs. business processes	Importance of specific processes
	Business processes vs. business issues	Effect of business issues on processes
	Business objectives vs. business processes	Importance of specific processes to objectives
Systems & technology	IT objectives versus architecture components	Applicability of objectives to architecture
	IT objectives versus IT issues	Impact of IT issues on IT objectives
	IT architecture versus potential technologies and systems	Potential benefits of new technologies and systems
	IT architecture versus IT issues	Impact of IT issues on architecture
Systems & business	Critical processes versus architecture components	Dependence of processes on specific architecture components
	Business issues versus IT issues	Alignment of issues
	Business issues versus architecture components	Impact of architecture problems on the business
	Critical processes versus IT issues	Impact of IT issues on processes
	Critical processes versus potential systems and technologies	Benefit of new technology on processes
Process group	Process group versus process group	Consistency and commonality
	Process group versus architecture	Dependence on common architecture components
	Process group versus business issues	Impact of business issues on processes
	Process group versus organizations	Involvement of departments in the processes
Individual transactions	Process steps versus process steps	Commonality
	Process steps versus infrastructure	Dependence and gaps in the infrastructure

Type	Improvement tables	Comments
	Process steps versus potential technology	Potential benefits to the steps
	Process steps versus process issues	Impact of issues on individual steps
	Process steps versus organization	Involvement of departments in individual steps
Current transactions	Transactions versus organizations	Involvement in entire transactions
	Transactions versus performance criteria	Comparative analysis
	Transactions versus business issues	Impact of business issues
	Transactions versus infrastructure	Impact of infrastructure
	Transactions versus process issues	Impact of process issues
	Transactions versus potential technologies and systems	Benefits of potential technologies
	Transactions versus IT issues	Impact of IT issues
Current process group	Process group versus business objectives	Support of business objectives.
	Process group versus business issues	Addressing of business issues
	Process group versus existing architecture	Impact of current architecture
Long-term transactions	Transactions versus organizations	Involvement in entire transactions
	Transactions versus performance criteria	Comparative analysis
	Transactions versus business issues	Impact of business issues
	Transactions versus infrastructure	Impact of infrastructure
	Transactions versus process issues	Impact of process issues
	Transactions versus potential technologies and systems	Benefits of potential technologies
	Transactions versus IT issues	Impact of IT issues
Future transactions	Transactions versus organizations	Involvement in entire transactions
	Transactions versus performance criteria	Comparative analysis
	Transactions versus business issues	Impact of business issues
	Transactions versus infrastructure	Impact of infrastructure
	Transactions versus process issues	Impact of process issues
	Transactions versus potential technologies and systems	Benefits of potential technologies
	Transactions versus IT issues	Impact of IT issues

Type	Improvement tables	Comments
	Transactions versus improved architecture	Fit with the new technology and systems
	Transactions versus process issues	Coverage of the process issues
	Transactions versus improved architecture	Fit with the new technology and systems
	Transactions versus process issues	Coverage of the process issues
Future process group	Process group versus business objectives	Support of business objectives.
	Process group versus business issues	Addressing of business issues
	Process group versus improved architecture	Reliance on new technology
Improvement implement. Strategy	Quick Wins in a phase versus business mission	
	Quick Wins in a phase versus business vision	
	Quick Wins in a phase versus business objectives	
	Quick Wins in a phase versus business issues	
	Quick Wins in a phase versus current processes	
	Quick Wins in a phase versus future processes	
	Quick Wins in a phase versus long-term processes	
	Quick Wins in a phase versus process issues	
	Future process changes versus business mission	Similar to that for Quick Wins
	Future process changes versus business vision	Similar to that for Quick Wins
	Future process changes versus business objectives	Similar to that for Quick Wins
	Future process changes versus business issues	Similar to that for Quick Wins
	Future process changes versus current processes	Similar to that for Quick Wins
	Future process changes versus long-term processes	Similar to that for Quick Wins
	Future process changes versus process issues	Similar to that for Quick Wins

APPENDIX 3

References

Breyfogle, F., *Implementing Six Sigma*. New York: J. Wiley and Sons, 1999.

Chowdhury, S., *The Power of Six Sigma*. Dearborn, MI: Dearborn Trade, 2001.

Eckes, G., *Making Six Sigma Last*. New York: J. Wiley and Sons, 2001.

Harrington, H. J., *Business Process Improvement*. American Society for Quality, Harrington, New York, 1991.

Lientz, B. P., and K. P. Rea, *Breakthrough Technology Project Management*, 2nd Edition. San Diego, CA: Academic Press, 2001.

Lientz, B. P., and K. P. Rea, *Project Management for the 21st Century*, 3rd Edition. San Diego, CA: Academic Press, 2001.

Pande, P.S., Neuman, R. P., Cavanagh, R. R., *The Six Sigma Way*. New York: McGraw-Hill Publishing, 2000.

Pyzdek, T., *The Six Sigma Handbook*. New York: McGraw-Hill Professional Publishing, 2000.

Tenner, A. R., and I. J. Detoro, *Process Redesign*. Englewood Cliffs, NJ: Prentice-Hall, 1996.

APPENDIX 4

Web Sites

Association for Work Process Improvement—www.tawpi.org
Process Improvement Laboratory—www.che.ufl.edu/pil
Department of Defense—www.c3iosd.mil/org/bpr.html
CERN—www.cern.ch/IPT
Association in Process Improvement—www.apiweb.org/API_hime_page.htm
Process improvement links—www.saigon.com/~nguyent/links.html
Six Sigma Benchmarking—www.sixsigmabenchmarking.com
General Electric—www.ge.com/sixsigma
Six Sigma Exchange—www.sixsigmaexchange.com
International Society of Six Sigma—www.isssp.org

INDEX

V

W